digital mcluhan

Alsager Library

Please return or renew by the date stamped below, or on recall

Telephone renewals: 0161 247 6130

WITHDRAWN

Do not mark this book in any way. A charge will be made for any damage.

"A ve

else, w

"Le

Always

works

right?' This may well be the most accessible introduction to McLuhan ever produced."

Review of Communication

“...explains why many of McLuhan's most memorable phrases – including 'The medium is the message' and 'the global village' – accurately anticipated today's digital society.”

Chronicle of Higher Education

“Essential for any thorough study of McLuhan's ideas in today's changing world.”

Midwest Book Review

“Levinson explains why Marshall McLuhan's theories about the media are more relevant in today's digital age than when they were first presented during the age of television.”

Library Journal

“Levinson seeks to understand the role of media in human affairs by re-examining the ideas of a man who died in 1980, at the dawn of CNN and more than a decade before Web pages. He makes an admirable case.”

Foreword Magazine

“Digital McLuhan is a very rewarding book.”

M/C Reviews

a guide to the information millennium

PAUL LEVINSON

London and New York

First published 1999
by Routledge
11 New Fetter Lane, London EC4P 4EE

Simultaneously published in the USA and Canada
by Routledge
29 West 35th Street, New York, NY 10001

First published in paperback 2001

Reprinted 2001

Routledge is an imprint of the Taylor & Francis Group

© 1999, 2001 Paul Levinson

Typeset in Perpetua by Routledge
Printed and bound in Great Britain by TJ International, Padstow, Cornwall

All rights reserved. No part of this book may be reprinted or reproduced or utilised in any form or by any electronic, mechanical, or other means, now known or hereafter invented, including photocopying and recording, or in any information storage or retrieval system, without permission in writing from the publishers.

British Library Cataloguing in Publication Data
A catalogue record for this book is available from the British Library

Library of Congress Cataloguing in Publication Data
Digital McLuhan: a guide to the information millennium/Paul Levinson.
Includes bibliographical references and index.
1. Mass media – information Revolution – Internet – technology.
2. McLuhan, Marshall. I. Title.
P90.L413 1999
98–40359
302.23–dc21
CIP

ISBN 0–415–19251–X (Hbk)
ISBN 0-415-24991-0 (Pbk)

MANCHESTER METROPOLITAN
WITHDRAWN
UNIVERSITY LIBRARY
G 302.23 MAC/LEV

TO JOHN M. CULKIN,

1928–1993

CONTENTS

"Rear-view mirror"

"Laws of media"

ACKNOWLEDGEMENTS

I first met Marshall McLuhan's work in 1964, as an undergraduate at the City College of New York. I would like to thank the professor who introduced me to *Understanding Media* (McLuhan, 1964) by name, but the truth is I cannot remember it – I was 17 at the time – and most of the good sense he showed in assigning that book to our class (I believe it was a seminar in educational psychology) was lost on me.

A decade would elapse before I encountered McLuhan's work in an academic setting again, but this time it was with all of the trimmings – not just a book, but McLuhan's work as a course of study – and I was a far more appreciative audience. That presentation of McLuhan's ideas was by John Culkin, who had arranged for McLuhan to spend a year at Fordham University in the late 1960s, and who by the 1970s had created an independent Center for Understanding Media, along with a Master of Arts in Media Studies offered first by Antioch College and then by the New School for Social Research, where I enrolled in 1974. John was a man whose personal warmth and zest for the knowledge he sought to convey was every bit as extraordinary as his comprehensive understanding of McLuhan, and my life as a scholar and media theorist would have been different indeed had I not had the good fortune to receive this kind of introduction to McLuhan from John. His death in 1993 prevents me from thanking him personally now, but this book, I hope, is testament to the send-off he gave to me.

By 1976, I had completed the MA in Media Studies at the New School, and had embarked on a Ph.D. in "Media Ecology," several blocks down the street in Greenwich Village at New York University. The guiding light of that doctoral program was and still is Neil Postman, who not only taught me how to teach, but conveyed valuable insights about McLuhan's work and its impact on the world. And Postman introduced me to more than McLuhan's work – he introduced me to Marshall McLuhan himself, a signal event in itself responsible for this book, and about which I will say more below. Neil Postman is of course an influential media theorist in his own right, and although I have had occasion to find fault with what I view as his unnecessarily pessimistic critique of media, his books are also cited in this book as noteworthy applications of McLuhan's thinking. Christine Nystrom and Terence Moran are other faculty in the "Media Ecology" Ph.D. program who helped teach me about McLuhan.

The people who were students with me in that doctoral program were as much a trove of enlightening conversation about McLuhan as were the faculty. Josh Meyrowitz and Ed Wachtel made the most prominent contributions. Their subsequent publications, like Postman's, are cited in this book.

By the time I was awarded the Ph.D. in 1979, two things had occurred in my life which had profound influence on my understanding of McLuhan, and perforce my writing of this book: McLuhan and I had become personal friends, and I was an Assistant Professor of Communications at Fairleigh Dickinson University.

The circumstances of my meeting with McLuhan, our brief but rich intellectual relationship in the few years before his death, are discussed at appropriate places in this book. Rather than previewing them here, I will simply state not only the obvious that McLuhan in his work is the person to whom I owe the biggest debt of gratitude for this book, but McLuhan in person was the most stimulating conversationist I have ever known. The dozen or so times we met – the chats we had over lunch, dinner, on the street, not to mention the numerous talks on the phone – were veritable alternate textbooks to McLuhan's published work, covering the same territory but often from a slightly different angle or altitude that revealed a camouflaged path to understanding.

As for my teaching, I must admit that, although I greatly prize this profession that is dedicated to the imparting of knowledge to others, I have always found a selfish motive for teaching as well: I truly learn something new, sometimes many things, in every class that I teach. My "Theories of Mass Communication" course at Fairleigh Dickinson University in the late 1970s and early 1980s was of course well stocked with McLuhan – his thinking dominated the class – and the very teaching of these ideas helped me clarify them. Likewise, the "Media Environments" course I taught in the early 1980s summer sessions at Fordham University's Graduate School were instructive about McLuhan – I hope to my students, but certainly to me.

My first course devoted entirely and explicitly to McLuhan was a "Seminar on McLuhan: 25 Years Later," that I created and began offering for the New School's MA in Media Studies (both online and in-person programs) in 1989. The "25 years" referred to the time that had passed since McLuhan's publication of *Understanding Media* in 1964, and I offered this course intermittently through the early 1990s with suitably updated titles. The student response to those courses corroborated what I already knew: that McLuhan's work was, if anything, increasing in relevance, almost as if our information age was crystallizing along the patterns of his vision. Undergraduate courses at Hofstra University and Fordham University in the middle and late 1990s gave me the same lesson.

I thus thank my students in every course I taught in which McLuhan's work played a role. Their questions and comments are among the unacknowledged catalysts of this book, so I recognize them here.

The two dozen essays, reviews, and books I have published since 1976 that deal in whole or in part with McLuhan's work have also been helpful in clarifying and developing my thinking about McLuhan, and I thank each and every editor who made those publications possible. Those texts are discussed in Chapter 1 and throughout this book, and are listed in the Bibliography.

I also thank Adrian Driscoll, editor of both my previous book, *The Soft Edge*, and this book. His encouragement was crucial.

Members of Marshall's family have been wonderful in the support they have given me, personal as well as intellectual, in my McLuhan studies over the years. Corinne, Marshall's wife, could not have made Tina (my wife) and me feel more welcome in their home during our trips to Toronto to see Marshall in the late 1970s. A dinner with Corinne, or even a chat, to this day easily remains the high point of any trip to Toronto, or indeed anyplace where I may be fortunate enough to meet her. Eric, one of Marshall's sons, was a big help to his father in his work on the "laws of media" near the end of his life. Eric's friendship, and his continuing application of his father's insights to the media of our day, has been a great help to me. Teri, one of Marshall's daughters, lives in New York City, which means we get a chance to meet a bit more often. In addition to her evocative work in anthropology, Teri is the keenest thinker I know about the place of her father in the intellectual firmament. Her counsel over the years has been invaluable.

I owe even more gratitude for this book to my own family. Tina Vozick and I were married a little over a year when we first went up to see Marshall in Toronto in the summer of 1977. One evening – after a day with Marshall at his University of Toronto office, and a superb dinner with Marshall and Corinne at their home – Corinne drove us to a nearby bus stop. But Tina and I were so excited by the day and its sparkling conversation that we could not stand still. So we walked, hand-in-hand for more than an hour, through the streets of Toronto to our hotel. Those streets truly seemed paved with magic that night, and indeed I walked them in every page that I wrote in this book.

Our children – Simon, 14, Molly, 11 – have been with us on that walk since they came into our lives. Authors often speak of "first readers" – people to whom they entrust a new manuscript for perusal before anyone else – but I have been fortunate indeed to also have a family of first listeners, discussants, and thinkers. Our conversations, increasingly enriched by insights and anecdotes and points of view provided by our children, wend their way through dinners, car rides, and vacations, and consequently through this book. In a sense, this book is more for their generation than ours, for they are, already, more fully denizens of the information millennium.

Paul Levinson
September, 1998

1

INTRODUCTION

Coinciding realms

Digital McLuhan is actually two intertwining books: one presents McLuhan's ideas about media and their impact upon our lives, the other presents my ideas about how McLuhan's ideas can help us make sense of our new digital age. I likely would have written a book like this in any case. But McLuhan could not, because he died on the last day of 1980, almost literally on the doorstep of the personal computer revolution that would change so much of our world, yet be so explicable via insights and comparisons McLuhan had earlier made.

Those insights showed us a world of media in motion, in which television was triumphing over books, newspapers, radio, and motion pictures for crucial segments of our attention, and consequently was exerting profound influence on politics, business, entertainment, education, and the general conduct of our lives. This Bayeux Tapestry of media in competition for our patronage – for our souls, according to some – quite naturally led McLuhan to consider the ways that media differed in their engagement of our mentalities. How and why, for example, does seeing a movie on television differ from seeing it in a motion picture theater, how is reading the news different from hearing it on radio, and how is that in turn different from watching it on TV? In raising and attempting to answer such questions, McLuhan in the 1950s, 60s, and 70s developed an intricate taxonomy of media and their effects, one which reached back to the origin of our species for comparisons – as in its recognition of similarities between pre-literate and electronic communication – and left openings aplenty for media to come. How is selecting the news we want to read, hear about, and watch on the Internet different from its presentation via newspapers, radio, and television?

The handwriting for coming to terms with our digital age was on the wall of McLuhan's books.

But that writing could be useful only to the extent that it was comprehensible – navigational clues, outlines of environments, in a language we do not clearly understand can serve to frustrate as much as educate. To the degree that McLuhan's taxonomy of media was cast in such a language, it was also a wall in the unintended sense of being an obstacle to understanding. McLuhan sought to call attention to the pre-eminent and overlooked role of the medium in communication – the difference between reading news in a newspaper and watching it on TV – with his famous aphorism, "the medium is the message." His critics and casual readers mistook that for a claim that the content – what it is we read in newspapers or watch on TV – is totally unimportant. He used provocative analogies to dramatize the differences between television and its competitors – cool versus hot, light-through versus light-on, acoustic versus visual space. But such metaphors often performed in ways precisely contrary to the reason that metaphors are employed: rather than elucidating a lesser-known area by relating it to an area that we know well, hot and cool and light-through and light-on were far more arcane than the media effects they sought to illuminate. When McLuhan was called upon to explain, he said his intention was not to explain, but to explore.

But this difficulty with McLuhan's presentation – with his medium of expression – in no way diminishes the importance of its content. It was there to see all along for those willing to learn the language, to see what was right with his game, as Wittgenstein advised we do about any novel ideas we may come across, in addition to subjecting them to criticism. Tom Wolfe asked in 1965, "What if he is right?" Might McLuhan be "the most important thinker since Newton, Darwin, Freud, Einstein, and Pavlov?" Although Wolfe surprisingly included Pavlov on that list, he was asking the right question. And the perspective of *Digital McLuhan* is that the answer is yes, at least insofar as a framework for understanding the human relationship with technology and therein the world and the cosmos is as important as frameworks for understanding the human psyche, life, and the cosmos in a physical sense.

Others came to the same conclusion. Critical anthologies such as Stearn's *McLuhan: Hot & Cool* (1967) and Rosenthal's *McLuhan: Pro and Con* (1968) contained as much admiration as denunciation of McLuhan (his language was so vivid, its claims so bold even when misunderstood, that inattention was rarely the first response). His biographers Philip Marchand (1989) and W. Terrence Gordon (1997) describe the caliber of protest at the University of Toronto's plan to close McLuhan's Centre for Culture and Technology in 1980, after he had been incapacitated by a stroke. Buckminster Fuller, John Cage, Woody Allen, Pierre Elliot Trudeau, and Jerry Brown were among the hundreds who called or wrote letters of support. The Centre was closed then nonetheless, but a recognition of McLuhan's importance was amply demonstrated at pinnacles of scholarly, cultural, and even political discourse.

In fairness to McLuhan's detractors, however, his work suffered from a problem more fundamental than the dazzle of his metaphors – a problem, moreover, not of McLuhan's making, and quite beyond anyone's capacity to redress at the time. In the scientific, and by extension social scientific, community, the surest way of determining to what extent someone's idea is right or wrong is to gauge the accuracy of its predictions. But there was no good occasion in the three decades McLuhan wrote about the media for either him or his associates to do this. From the publication of *The Mechanical Bride*, his first book about media, in 1951, to his death on the eve of 1981, television utterly dominated public and private life. This meant that any predictions McLuhan made about television, or were generated by others from his ideas, had elements of the ex post facto, of explanations of an environment already present. Thus, McLuhan observed that JFK's "cool" style was more appropriate to the medium of television than Nixon's "hot" arguments in 1960, and Jimmy Carter's low-key persona similarly made him appealing to voters making decisions on the basis of television sixteen years later. But as intriguing and useful as those connections may have been (and still are), they could not provide the corroboration of McLuhan's ideas that the arrival of a new medium – as revolutionary and unforeseen in its impact as television – might have bestowed, had the advent of that new medium and its effects been predictable and explicable on the basis of McLuhan's work.

The digital age now provides such an occasion. *Digital McLuhan* thus not only seeks to provide a guide to our digital age – the book's primary purpose – but in so doing provides evidence of the underlying accuracy of McLuhan's thinking that was unavailable when he was alive.

To accomplish this double task, each chapter of the book will attempt to clarify a key insight, principle, or construct in McLuhan's work, and then discover what it tells us about tens of millions of people reading the Sunday papers, purchasing birthday presents, and even watching the equivalent of television on the Web, while those not online learn about it anyway in newspapers, magazines, movies, and TV shows.

THE GAME PLAN

We begin in the next chapter with a consideration of McLuhan's method – his professed preference for exploration over explanation, for demonstration via metaphor rather than logical argument, and his presentation of ideas about media in small packets, often as few as several paragraphs and rarely more than seven or eight pages. Strictly speaking, this is not about an insight into media effects nor a tool or construct McLuhan developed for assessing them. Rather, it is about McLuhan's way of doing business with his readers – his modus operandus. And yet not coincidentally, this turns out to bear a striking similarity to the way people communicate online, with comments on Usenet lists usually just a few paragraphs in length, and hot-linked titles and phrases on Web pages much like the glosses in bold – hot-links ahead of their medium – we find dispersed throughout the pages of McLuhan's books. Even in the examination of his method, we find in McLuhan a presaging of our age, an author struggling to communicate in an electronic pattern via the straightjacket of paper – a startling quicksilver mode in some sense consonant with the wheels of our intellect, but not as yet invented then in media.

Next, we turn to the core insight of McLuhan's entire agenda, also his best known and least understood: the medium is the message. Intended to call attention to the proposition that mere use of a medium is more profound in impact upon society than what individuals may do specifically with the medium – the world successively changed when people began talking on the phone, listening to the radio, watching television, logging on to the Web, not usually because of what they said, heard, and saw – it was roundly mangled into a contention that content is totally unimportant.

A moment's reflection shows why that cannot be. There is no such thing as a medium without content, for if it had no content, it would not be a medium. McLuhan (1964, pp. 23–4) cites the electric light as a hypothetical example of "pure information," or a medium without content, but then aptly notes that its content is what it shines upon and illuminates. In other words, the light bulb becomes meaningful to the extent that it illuminates something. A television with no programs could have no influence upon us as a medium, any more than a computer devoid of its very different kind of programs would be anything other than an interesting piece of junk. This is, indeed, just what became of many early personal computers due to their inability to access the Web, as soon as such access became crucial. They lacked the programs, and thus the content the programs delivered, necessary for the computer to function as a medium in the new environment. Content, in other words, is essential for "media-hood."

The Internet highlights another way in which the content of the medium aids our understanding of media. In McLuhan's quest to uncover the ordinarily hidden dimensions and effects of media – unnoticed because we focus on content and take the underlying medium for granted – he observed that media suddenly become more visible and attractive as objects of study when they are superseded by newer media, and become their new content. Thus, McLuhan's early work in literary theory showed him that the narrative structure of the novel jumped into public awareness after motion pictures adopted that structure as its content. By the 1960s, television would have the same effect on cinema, as universities created film schools to examine what was now available as content at all hours of the day in everyone's home. And in the decade after McLuhan's death, the VCR transformed the very structure and organization of television into content for the first time, directing the attention of its viewers to the relationship of commercials and programming (commercials could be "fast-forwarded" on the VCR), the subtleties of program timing (the taping could end several minutes before the end of the program, because the rest was taken up by commercials), and other aspects of television uncritically accepted when they were beyond viewer control.

But in the new millennium, the Internet is poised to trump each and every one of these prior "liberations" of media into content, because the Internet is making content of them all. What began as a medium whose content was text, and expanded in the 1990s to include images and sounds, has become at the turn of the new century a medium that offers telephone (Internet Telephone), radio (RealAudio) and television (RealVideo). The evidence and implications of the Internet as this grand medium of media will be among the continuing themes of this book.

Chapters 4 and 5 address McLuhan's discussions of "acoustic space" and "discarnate man," and therein consider the impact that the Internet as a whole has on our relationship to the world, and to one another. A prime concern of McLuhan's was the way the alphabet and the printing press encouraged us to see the world as a series of discrete sources and pieces, from which we could be easily detached, as when closing a book. According to McLuhan, such abstract, sequential vision had replaced an earlier, "acoustic" mode, wherein we perceived the world all at once, all around us, as a permeable extension of ourselves and we of it. Provocatively, McLuhan claimed that television was retrieving this mode – via screens that showed the same thing everywhere we turned. But regarding television as "acoustic" was a difficult feat, no matter how often McLuhan aptly quoted Tony Schwartz (1973) that television treats the eye as an ear.

The advent of cyberspace in the 1990s made it easier.

For the space that the computer screen invites us to join is indeed everywhere, but unlike the space on the television screen, it is potentially of our own making – we create it and remake it by using it – just like the acoustic space of the pre-literate environment. Further, the notion of being *in* cyberspace is much less counter-intuitive than being in the acoustic space of television. We go from one place to another on the Web and we feel as if we are moving through that space – a sense we do not usually have when jumping from one television station to another. The unmasking of cyberspace as acoustic space thus helps make each more explicable.

Denizens of cyberspace are virtual – meaning, our physical bodies play no role in our interaction with and in that space. McLuhan noted this "discarnate" effect when we talk on the phone, listen to the radio, or watch television, and wondered what impact it had upon our morality. But the experience on the phone is very different from the other two, in that we become discarnate when we talk on the phone – each party to the conversation is "sent," without accompanying body, in every word that is spoken – but only the viewed, not the viewer, is discarnate on television. The online participant is incorporeal in the same interactive way as the telephone conversationist, and on this key aspect the Internet is more like the telephone than television. Indeed, we will see throughout this book that the digital age has tap-roots in telephones and printing every bit as powerful as those in television, even though the digital age is brought to us on screens first made familiar on TVs.

Chapters 6 and 7 focus on the geo-political consequences of this revolution. McLuhan suggested that electronic media, television in particular, were turning the world into a "global village." The logic of this observation is immediately comprehensible and made the global village not only among the most often but also the most appropriately quoted of McLuhan's metaphors: we can all see the similarity between the world watching the Superbowl on television and villagers enjoying a local football game from their vantage point in a stadium. But there is more to village life than being a passive audience – villagers in a stadium can interact with one another and the players, and indeed the players may be villagers themselves – in contrast to the television audience, which consists for the most part of isolated family units, at irreducible arm's length from what they watch on the screen. Once again, the Internet helps complete McLuhan's metaphor, to the point of making it a reality. The online villager, who can live anywhere in the world with a personal computer, a telephone line, and a Web browser, can engage in dialogue, seek out rather than merely receive news stories, and in general exchange information across the globe much like the inhabitants of any village or stadium. Just as D. W. Griffith shattered the shell of the proscenium arch that had kept movie cameras, still under the spell of theater, from approaching the scene too closely, so has the Internet shattered the barrier that kept viewers bottled up with no input on the living-room side of their television screen.

The advent of computer screens not only as receivers but initiators of information in homes and offices around the world is further fulfilling another of McLuhan's observations about the global village – namely, that its dispersion of information is creating a new power structure whose "centers are everywhere and margins are nowhere." Radio and television networks began this process by broadcasting the same breaking news into everyone's homes and offices, and even motel rooms. From the point of view of access to this important information, the room with the best view could just as easily be in a shack off a desolate road in the middle of nowhere as a penthouse or office suite in New York City – all that counted was that the room have a television or radio (in a sense, this effect began with national news magazines, although their delivery was not immediate). But the sources of this information were still controlled by a handful of broadcast networks; in the television age, their corporate headquarters were meaningful centers indeed.

In the age of the Internet, in which anyone with a Web page can launch a news story, internationally, the corporate gatekeeping of news is finally beginning to subside. I learned about Princess Diana's tragic accident in August 1997 via an Associated Press bulletin forwarded by an individual on the Internet. Although some of the cable television stations were quick to pick up the story, an hour or more elapsed before all the major U.S. television networks joined in the coverage. Similarly, the Starr Report on Bill Clinton's sexual relationship with Monica Lewinsky (Office of the Independent Counsel, 1998) was available in its entirety to the world at large on the Internet at a time when only excerpts were quoted on radio and TV, and a day before its publication in newspapers.

The decentralization of our digital age pertains to more than news. Amazon.com became the third largest bookseller in the world within the first three years of its online operation (Nee, 1998). There is of course a centralized corporate structure in Amazon.com, but it is irrelevant in terms of the books that are offered for sale to its customers: unlike even the biggest physical bookstore, which can only shelve a given number of different books, the shelf space on Amazon.com – being virtual – is virtually unlimited.

And in many cases, the power of corporations to influence economic events, much like the power of governments to influence them, is melting in the light of personal computers and their empowerment of individual choice. Microsoft, the biggest corporation in the world, was unable to make its Windows 95 a thorough success, just as it has been struggling for years to attain a majority of the market for its Web browser, Internet Explorer: in both cases, the preferences of individual users, not the plans of the mega-corporation, prevailed. That is why lawsuits by the government to limit the power of Microsoft are unnecessary. The power is already limited by a decentralization, in significant part of Microsoft's own making, far more profound. Indeed, for anyone who understands McLuhan, the lawsuits are laughable – a quixotic effort by government, eager to demonstrate what little power it has left to regulate commerce, against an alleged monopoly whose very success has had the effect of obsolescing monopolies in the information business. Call it a tilting at Web mills.

McLuhan's examination of media looked not only at their impact on business, politics, and social life, but at the way they address and engage our senses – the locus of their primary psychological effect upon us, which is the basis of much of their social impact. Chapters 8 and 9 turn from our consideration of the global lessons of the digital age to a new look at the one-on-one relationship each of us has with our computer screen, our television screen, our books, our media.

For McLuhan, the specific way that we perceive the contents of each medium – the literal, physical method that each medium employs, the intensity and clarity of the information therein presented – governs not only how we use and what we derive from the medium, but its effect upon our overall society. Observing a common perceptual denominator in television and stained glass windows – both are animated by light behind the glass, which reaches our eyes after shining through it – McLuhan came to an astonishing conclusion about television: it draws and commands our attention with an almost hypnotic, religious intensity, because that is the way our senses and brains respond to a "light-through" invitation. Stained-glass windows – and, I would add, blue skies – partake of this sensory appeal. Paintings, books (other than those comprised of illuminated manuscripts, the medieval simulation on pages of light-through), newspapers, and motion pictures do not; rather, their contents are conveyed to us via light that bounces on and off of them. Thus, on this sensory basis alone, we can see the advantage of television over both books and motion pictures. Since computer screens also operate via light-through, they convey the benefits of books while maintaining the sensory appeal of TV.

"Light-through" is likely among the least known of McLuhan's comparative gauges of media. Like "acoustic space" and "discarnate man," it is appreciated among some inner circles of media theorists, but never achieved the iconic status of "the global village" or the mantric appeal of "the medium is the message." "Hot and cool" – a sensory comparison that McLuhan borrowed from jazz slang to denote media of high and low profile – went its own different way. An early star in the mid-1960s, at least as well known and associated with McLuhan as the global village and the medium is the message, it fell sharply out of style after McLuhan's death in 1980, and today has about it an almost charming, antique patina. Its one consistency throughout the decades is the large amount of misunderstanding it has generated – which, unlike the medium is the message, is due to some inherent difficulties, or unannounced subtleties, in the concept.

The crux of hot and cool is that media which are loud, bright, clear, fixed ("hot" or high definition) evoke less involvement from perceivers than media whose presentations are soft, shadowy, blurred, and changeable ("cool" or low definition). The psychological logic of this distinction is that we are obliged and seduced to work harder – get more involved – to fill in the gaps with the lower profile, less complete media. Thus, we might pore over a few lines of poetry more than a few lines of prose, study a political cartoon for meaning more than a crystal-clear photograph, get hooked on the shimmering images of the little screen of television more than the big bold images in the movie theater.

That last example shows the strengths and weaknesses of the hot-and-cool dichotomy. The revelation that television pulls us into its cool images to see what is on the other side dovetails nicely with the same effect that television has upon us as a light-through medium. It is true that TV is small and blurry in comparison to the motion picture screen, which is thus not only light-on but hot in this McLuhanesque reading. But when we inquire more fully into why TV might be more involving than motion pictures, we notice that the two differ on another crucial criterion – namely, that television is available to us twenty-four hours a day in our homes, whereas motion pictures on the theater screen can be had only by going out to the movies and paying for the experience. This difference alone could easily account for how television could be more addictive than motion pictures, with no reference to hot and cool, or light-on versus light-through, at all.

Other non-perceptual factors can account for hot-and-cool effects of other media. McLuhan correctly noted that radio and phonograph recordings ("hi-fi" or "high-fidelity" in the late 1950s and early 1960s) provided much fuller sound than the telephone, which was thus cool and much more involving. But the telephone invited participation for a much more practical and obvious reason: unlike the radio and the phonograph, the phone provided the listener with a live, interacting person on the other end of the connection. Whatever the heat or coolness of radio and recordings, they could never warrant the level of involvement of telephone, for the plain reason that they are deaf to the voices of their listeners (unless, of course, someone calls a radio station on the phone – in which case, the addition of the phone makes radio interactive).

And radio poses another problem for hot/cool analysis. How is it that radio, a sound-only medium, can be hot, while television with its audio-visual presentation can be cool? Surely sound is less than sight-and-sound? We can answer this one with the qualification that hot and cool works best for media of the same modality – motion pictures and TV, prose and poetry, cartoon drawing and photograph – and not for cross-media comparisons such as radio and TV. But we are still left with the unavoidable result that hot and cool is a rather mercurial measure of media effects, which (perhaps apropos of the somewhat disenchanting spin it puts on clarity) can confuse as much as clarify.

Nonetheless, its elucidations are impressive, and it and they are included in this book because we need all the help we can get in furthering our understanding of the new digital age. Text online has been more addictive than books and newspapers ever since its public inception in the 1980s, when early commercial systems such as CompuServe and the Source would charge by the hour, and people who could ill afford the cost would spend a thousand dollars or more a month and still continue logging on. The interactivity – sometimes live, usually asynchronous – with other people online certainly explains a large part of the attraction. But there was something in those primordial, low-profile screens – just letters of one shade on a dark background, dim and drab in comparison to the bright screens of today – that was powerfully alluring, and made me think of McLuhan and cool media writ in phosphor the first time I logged on to a computer network, almost fifteen years ago.

To write is usually to want to publish – a diary would be the exception – and in Chapter 10 we return to the social arena to consider what the Web has made of McLuhan's proposition that the xerox was turning every author into a publisher. The initial formulation of course had more than a dash of hyperbole. Even the cheapest paperback looks and feels more like a published book than the sharpest photocopied manuscript, and the author with photocopied "publications" in hand has no ready way of getting them into public distribution. The Web is a great equalizer on both accounts. Authors with the requisite minimal knowledge of HyperText Markup Language (html) and sense of Web-page design can create online pages for their publications as attractive as those put up by the biggest corporations online, and the Web is a universal distribution system whose pages can be accessed by anyone with a Web browser. But this revolution in publishing is still decidedly incomplete: Amazon.com's huge business is in selling, via the Web, traditionally published books.

Whether the buoyance of the book printed and bound is an expression of the nostalgia all of us who have grown up with it feel for it, or an indication of deeper levels of satisfaction we derive from words permanently affixed to pages – *Digital McLuhan* will examine both factors, and I suspect both are at work here – the currents of change swirling around books and newspapers are all headed in one overall direction: a washing away, an overwhelming, of traditional gatekeeping in media. When means of dissemination were handwritten and thus at their scarcest, the Church served as gatekeeper for sacred and lesser-blessed but still worthy text. The printing press took down this gate, but installed in its stead the government and soon commercial enterprises to regulate the new flow of information. In the twentieth century, the broadcast media of radio and television radically increased the flow of information again, yet government and commercial gatekeeping continued unabated – indeed, even increased in authority, since publishing a book is considerably less expensive than producing a television program, which until the advent of cable provided only a handful of possibilities for appearing on television screens in any case. McLuhan was thus entirely right, even prescient, to seize upon the xerox as an extraordinary reversal of this trend. The question for gatekeeping in the digital age will be: with the Web removing the technological and economic reasons for the pre-sorting of information, will the public still look to gatekeepers to provide an imprimatur of what is best to read, see, and hear, or will audiences seek out and ratify a more direct relationship with creators?

Chapters 11, 12, and 13 consider how digital facility with information may be changing our very notion of "best," and how it relates to our interconnected conceptions of work, play, and art. McLuhan not only had a genius for apt phrases, he had a keen eye for picking them out from the discourse of others, and frequently quoted the saying of the Balinese that "We have no art, we do everything well." Along with acoustic space and the global village, McLuhan saw this pre-industrial attention to detail, this goal of working to perfection, as returning in an electronic age in which expert knowledge was increasingly available to everyone. As has been happening in so many other aspects of our lives, the personal digital age has done one better than the mass electronic age in this regard, giving us not only access to information twenty-four hours per day but the wherewithal to apply that information, to contribute to society, to even work at a growing number of jobs any time night or day, from any place in the world, including our homes.

The source of this enhanced capacity for work is of course the personal computer, also a place where our children (and we) play games, and via which we surf the Web for fun as well as profit. "Serfs to Surf" (Chapter 11 of this book) examines some of the background and likely consequences of this new, digital blurring of work and play. Although television in its recently departed classic age no doubt mixed entertainment with news and business (in the form of commercials), it provided few outlets for direct purchase of goods and services, and none at all for production and distribution of work from the home; business in those days by and large could only be conducted outside of the home, in physical public. In contrast to such sharp demarcation, the personal computer from the outset was a vehicle both of work (word processing, data management, telecommuting) and pleasure. Indeed, the polarity of DOS (business) and Macintosh (fun) applications captured this double function, whose ultimate synthesis in Windows was inevitable, given that the actual differences between DOS and Mac programs were minimal in comparison to personal computers versus any media that came before. Whether the fun of surfing the Web will continue when the novelty wears off, whether the work produced in such flexible circumstances will be better in the long run, whether family life will be improved in all respects by the capacity to work at home, is not yet known. But the robust levels of economic growth in the United States at the end of the 1990s – rising GNP, declining unemployment – suggest that, in the most Web-connected nation in the world, at least, the new mixture of work and play is producing a fine harvest.

Chapter 12 looks at the other end of work and pleasure in relation to technological progress, particularly McLuhan's view that outmoded technologies become art forms. As a corollary to his warning that technologies are essentially invisible when in peak use – we might say that they are all like the blades of whirring fans, upon which the unwary might well cut their fingers – McLuhan noticed that technological workings suddenly become clear, as if pushed onto center stage, when another technology takes over even part of their job and begins pulling some of the strings. One aspect of this, as we saw in our discussion of "the medium is the message," is that older media become the high profile content of newer media, as novels in motion pictures, motion pictures in TV, and almost all prior media on the Internet.

Another aspect is that some superseded technologies become appreciated not for their actual output or function, but for the sheer pleasure of experiencing them – as we would look back at and enjoy a work of art. McLuhan's favorite example of this effect concerned the Earth itself, which became an art form – a thing of beauty to be admired and preserved as a whole, as if Gaia were an endangered species – when Sputnik circled our planet, and for the first time gave us a perspective from beyond it. My own two favorite examples (a key enjoyment in reading McLuhan's work is coming up with your own examples) are delicatessen meats and being cool in convertible cars. Ham, corned beef, and similarly treated food were once cured for the practical purpose of preservation; when electric refrigeration was developed to do a much better job of that, people began to consume cured foods almost entirely for their taste. Around that very same time, people drove in convertibles in summer to keep cool; by the 1960s, air-conditioning in cars had all but eradicated the convertible; in the

1980s, the convertible returned – its occupants wanted to be cool, but now in a stylistic more than a physical sense.

As medium after medium moves from its traditional stand-alone position to become content on the Internet, we can expect a commensurate increase in the public's appreciation of those older media as art. The look and design of print on paper is already beginning to receive such attention – see the front-page layout of any newspaper in comparison to what front pages looked like a hundred years ago – just as handwriting became the art of calligraphy in the age of print. The appearance of television on screens attached to keyboards and trackballs will make it not only newly viewable but malleable in its contours, and this invitation to experiment with the size and shape of the TV screen, and its relation to other windows, will result in an increased awareness of what are now the underlying aesthetics and structures of the televised image. A small first step in that direction has already occurred in the repackaging of 1950s TV situation comedies – such as *I Love Lucy* and *The Honeymooners* – for presentation on cable television stations as "classics," replete with commentary at the beginning about their comedic significance. The availability of nearly 100 cable channels in many areas of the United States constitutes an enormous increase in choice over the handful of broadcast channels which was all that television had to offer in the 1950s, but the Internet has the potential to offer vastly more, and may soon make even cable TV an "Internet, lite."

But will this increase in art à la McLuhan – this shift in many older technologies from our unthinking use to our critical appreciation of them – result in a net improvement of society? In other words, granting that the Internet does indeed afford us more time and opportunity for the careful attention to activities typical of art, will this translate into the Balinese supposition of doing all of that well? Chapter 13 considers some of the prospects and pitfalls for genuine improvement of our ways of life and business in the digital age. On the one hand, rapid access to diverse information certainly facilitates better performance in any task that requires research, which cuts a wide path across medical, legal, academic, and many commercial professions. On the other hand, there can be a sense of accomplishment and relationship in the virtual realm which is illusory, or at least incomplete, until full flesh-and-blood people have shaken hands, and tangible things have been moved.

To the extent that pre-industrial Balinese were indeed able to do everything well – given the deception attendant to most self-appraisal – this presumably was due to the smaller number of tasks in their world, and the consequently greater amount of time they could devote to each, in comparison to less time for more tasks in the Industrial Age. Digital processing of data effectively increases the amount of time we can devote to each task when it provides information essential to the task more quickly; and successful completion of more tasks gives us access to the bigger picture of interrelated tasks, which is also helpful. So far, so good: we seem well on the way to Bali Ha'i.

But the irreducible grounding of many tasks in physical reality and their refractory time frames – the Internet, after all, cannot move an orange from California to London any faster than the fastest plane, nor can the Web do anything about the time it takes an orange to grow – is an ever-present anchor on the digital surge, ever ready to bring it back down to Earth. In the end, we may well need to settle not for doing everything well, but for some of us doing some things better – which isn't too bad, either.

Peering into the future brings us to the last two chapters in this book, where we find McLuhan segueing from tour guide for our world – a world twenty years after his demise – to docent of further worlds beyond. Consistent with McLuhan's abstention from explanation and overarching theorization, he also refrained from systematic, detailed predictions of the future. His forte was rather the sudden dive into the past, to retrieve some sparkling gem from the ocean deep to help illuminate our present state of affairs back up on the surface. Thus, we have the global village, acoustic space, stained glass windows all shedding light, first on McLuhan's world of the 1950s through the 1980s, and now on the digital age in which we reside, an age which is just beginning. But McLuhan also left us, if not explicit predictions of the future, two very valuable conceptual tools to help us navigate its pathways. One, the "rear-view mirror," is designed to alert us to errors in perception, traps we may fall into, along the way. The other, the "tetrad" or "laws of media," was intended to uncover panoplies of possibilities, and how they relate to past and current effects of media.

The rear-view mirror – subject of Chapter 14 – is, like the global village, among McLuhan's easiest to understand and most powerful insights. In fact, the rear-view mirror is probably my favorite. We move into the future with our sight on the past. How true that is, and how aptly the rear-view mirror in the automobile metaphorically captures that effect. Its linguistic traces are all around. The telephone was first called the talking telegraph, the automobile the horseless carriage, the radio the wireless. But each of these technologies was much more – the telephone breached the privacy of our home, the automobile empowered countries which had oil, radio became a nationwide simultaneous mass medium – and since none of these consequences were picked up in the initial retro-labels, those rear-view mirrors distracted us from crucial developments.

The Internet, of course, is seen in a rear-view mirror par excellence. Its critics are prone to see it as a television screen; its devotees, including me, are inclined to see it as an improved kind of book. But the truth of the matter, yet to be fully determined, is that the Internet is and will be a combination and transformation of both books and TV, and other media such as telephone as well, and thus is something much more, much different from any prior media. The rear-view mirror cannot tell us what that is, but it can remind us not to get too mesmerized by reflections of the immediate past. The driver who looks only into the rear-view mirror, or even too often, and accords consequently short shrift to the road ahead and its new possibilities can quickly end up on the side of the road, or worse.

McLuhan's "tetrad" – or four laws of media – shifts gears from warning us that we need to sometimes take off our past-tinted glasses when looking at the future, to indicating what kinds of territory we might see there when those glasses are removed. McLuhan says four questions can be asked of any medium and its impact. What does it enhance or amplify in the culture? What does it obsolesce or push out of prominence? What does it retrieve from the past, from the realm of the previously obsolesced? And – and here the tetrad projects into the future – what does the medium reverse or flip into when it reaches the limits of its potential? Radio, for example, enhanced oral communication across great distances; obsolesced aspects of written communication, such as the newspaper as the leading edge of news delivery; retrieved some of the prominence of oral communication from pre-literate times; and reversed into broadcasts of sounds and images – television. The new medium can in turn be similarly examined. Television enhances instant visual long-distance communication; obsolesces aspects of sound-only radio, whose serials and soap operas moved to TV; retrieves some of the visual elements that radio had obsolesced, such as the cartoon; and reverses into.... Well, we are not entirely there yet, but certainly oligarchic network television has reversed into such diverse but overlapping media as cable, VCRs, and the Internet.

Chapter 15, the final chapter in our voyage, applies the tetrad to these new media of our early digital age, and looks especially at their possible areas of reversal. Might the Internet, with its democratization of communication, its scoffing at gatekeeping, reverse in the next turn of its tetrad into a Web in which choice is a sham, where every hot-link on a page leads to a pre-determined info-dump, mandated by the government or some other resurgent central authority as in the movie *Starship Troopers* (1997)? Or are the centrifugal forces that the digital age has already unleashed so powerful, so quickly on their way to becoming universal, that they will pull the tetrad wheel away from any re-entry of a totalitarian past?

The tetrad is unique in Dr. McLuhan's bag of instruments to gauge the health, status, heartbeat, prognosis of our media. Although it is more systematic than any of his other constructs – every medium in history is subject to the four laws – it is nonetheless open-ended and multi-dimensional. Each medium enhances, obsolesces, retrieves, reverses into many more than one thing or effect. Television retrieves the cave painting, the billboard, the newspaper cartoon, and reverses not only into the Internet, cable, and the VCR but holography, the videophone (which of course is also one of the many media the telephone flips into) and much more. Clearly, McLuhan was attempting to do something new with the tetrad – to create something which approached a general theory of media, without the rigidity and confinement of possibilities which are usually the purchase price of theories. But he never got the chance.

He managed only two small, published essays about the tetrad before his death in 1980 (McLuhan, 1975, 1977a). The laws of media were not able to make their first house-call to the world-at-large until 1988, when the book by that name, co-authored by Marshall and Eric McLuhan, was at last published.

The tetrad also has a personal uniqueness to me: I wrote the Preface to the second article McLuhan published about the tetrad – "Laws of the Media" in the journal *et cetera* in 1977. I was a doctoral student then, and although I had already studied McLuhan's work at length, that Preface initiated a brief, exhilarating period of time in which I not only read McLuhan's work, but worked with him through correspondence, phone calls, and the occasional meeting in Toronto or New York.

The tetrad thus not only serves as the culmination of this book, but provides apt occasion for the conclusion of this chapter, where we will situate Marshall McLuhan in relation to his colleagues, co-authors, and those in the scholarly world who continue to further his work.

McLUHAN & COMPANY

One indication of the extent to which McLuhan relied on co-authors is that only three of his books – *The Mechanical Bride* (1951), *The Gutenberg Galaxy* (1962), and *Understanding Media* (1964) – were written without collaboration. All of the others – *The Medium is the Massage* (McLuhan & Fiore, 1967), *War and Peace in the Global Village* (McLuhan & Fiore, 1968), *Through the Vanishing Point* (McLuhan & Parker, 1968), *Counterblast* (McLuhan & Parker, 1969), *From Cliché to Archetype* (McLuhan & Watson, 1970), *Take Today* (McLuhan & Nevitt, 1972), *City as Classroom* (M. McLuhan, Hutchon & E. McLuhan, 1977), and the posthumously published *Laws of Media* (M. McLuhan & E. McLuhan, 1988) and *The Global Village* (McLuhan & Powers, 1989) – were co-written with the authors indicated in parenthesis. (*Culture is Our Business*, written solely by McLuhan and published in 1970, is essentially an update of *The Mechanical Bride*; *Verbi-Voco-Visual Explorations* is a revised 1967 reprint of number 8 of the journal *Explorations*, which McLuhan edited with Edmund Carpenter; McLuhan and Carpenter also edited *Explorations in Communication* in 1960.)

A counter-indication of McLuhan's indebtedness to co-authors, however, becomes clear when we realize that *The Gutenberg Galaxy* and *Understanding Media* – two of McLuhan's three solo works – have been far and away his most influential books. Indeed, examination of his subsequent books reveals that, although they continue to offer some new examples, the insights and metaphors and media relationships they illuminate – such as the global village, hot and cool, light-through versus light-on, and, of course, the medium is the message – are all already writ large in his second and third books. *Laws of Media* is a partial exception, but retrieval was already on stage in the global village, and reversal was the subject of a chapter – "Reversal of the Overheated Medium" – in *Understanding Media*. Enhancement and obsolescence – the replacement of one medium in prominence by another – are obviously major themes throughout McLuhan's work, in particular the obsolescence of the oral by the literate, and then the literate by the electronic. *Digital McLuhan* can be seen as an examination of the obsolescence of the early, mass-electronic milieu by our current electronic digital environment – a replacement which both enhances the world-wide reach of earlier mass electronic media, yet retrieves literate culture via the reliance of the personal computer and its Web applications on the written word. What was unapparent when McLuhan and all but his posthumous collaborators wrote was that the electronic media of their time were but the root of an electronic digital age on the verge of springing into being – a one-way, mass-media caterpillar yet to burst into the digital butterfly, if you prefer your metaphors winged.

Since the presentation of McLuhan's co-authored books makes it impossible to determine which of the authors wrote what text in a given volume – and since, as indicated above, the insights in the co-authored texts are found in the earlier books solely of McLuhan's authorship – I will, for brevity's sake, refer to quotes and ideas discussed throughout this book as "McLuhan's," while of course listing any co-authors in the bibliographic reference immediately following the quote (for example: McLuhan & Fiore, 1967).

Books solely written by McLuhan's collaborators, or books written about McLuhan and his ideas, are of course a different matter. Yet, significantly, only one book by a McLuhan colleague – Edmund Carpenter's *Oh, What a Blow that Phantom Gave Me!* (1972/1973) – stands out as making a contribution essential to our understanding of the digital age, and indeed that will be the only book of its kind quoted in this volume. As for early critical works about McLuhan, they are alas usually distinguished by a stark misunderstanding of their subject – see, for example, Jonathan Miller's *Marshall McLuhan* (1971) and its deadpan explanation that McLuhan applies "hot" to incomplete, low-profile media, or completely the reverse of McLuhan's usage – and will be cited in this volume as such.

Beyond McLuhan's collaborators and critics, we have a small but growing group of authors who have applied and extended McLuhan's work to new areas. Neil Postman (e.g., 1985, 1992) has mined the dark side of McLuhan, contending that regardless of how newsworthy, educational, and otherwise culturally erudite the content of television may appear, its underlying message – the result of watching it, whatever the program may be – is destructive of rationality, courtesy, and many of the finer things of civilization. Jim Curtis' *Rock Eras* (1987) applies the tetrad to rock 'n' roll. Joshua Meyrowitz's *No Sense of Place* (1985) develops the social and political implications of McLuhan's observation that life at electronic speed blurs distinctions between work and art, business and pleasure.

Much of this scholarship was interconnected, and informed by personal relationships among authors that continue to this day. I first became acquainted with Jim Curtis when I reviewed his *Culture as Polyphony* (1978) for the journal *Technology and Culture* (Levinson, 1979b); I noted his analysis of Russian and American Southern culture as "cool" and acoustic, and we struck up a correspondence and a friendship. His *Rock Eras* (1987) not only uses McLuhan's tetrad, but a particular elaboration of the tetrad that I developed – tetrad "wheels" of evolution – which is discussed in Chapter 15 of *Digital McLuhan*.

Josh Meyrowitz sat right next to me in seminars of the doctoral program at New York University – the "Media Ecology" Program, headed by Neil Postman – from 1975 through 1978, where our relationship rapidly developed from spirited sparring over minor details to lifelong friendship and a recognition that we both saw the world of media essentially the same way. That core vision derived from Marshall McLuhan. I went on in the years that followed to integrate that vision with the philosophy of Karl Popper, the evolutionary epistemology of Donald Campbell, and even some of the logical analysis of technological possibilities found in the science fiction of Isaac Asimov. Meyrowitz mixed his McLuhan with the perspectives of sociology, most especially the work on public and private personas by Erving Goffman. Although *No Sense of Place* (1985) was just a bit too early to catch the full drift of the digital revolution – the book is much more an assessment of television than computers – its comprehensive treatment of fading distinctions and boundaries in electronic mass media offers an important prelude to the frontal assault on gatekeeping in the digital age that is one of the main themes of *Digital McLuhan*.

Neil Postman was not only the intellectual leader of the program in which I earned my Ph.D., he was also my doctoral dissertation adviser. In view of his scathing critiques of television and computers, and my opposing arguments that they are both far better for our culture than Postman allows, and radically different in any case in their effects, I have often been moved to quip that Postman was my biggest failure in which I was a teacher – in this case, a doctoral student attempting to "teach" his mentor about media. But the deeper truth is that Postman was and is a superb teacher in conveying why McLuhan must be taken seriously.

Indeed, Postman and his doctoral program were responsible for several crucial stages in the intellectual and personal relationship I came to enjoy with McLuhan, especially its beginnings. My first published scholarly article – in the *Media Ecology Review*, a publication of the Media Ecology Program at NYU – was entitled "'Hot' and 'Cool' Redefined for Interactive Media" (1976), and its thesis that hot and cool distinctions work differently in passive one-way media like radio and interactive two-way media such as telephone points directly to new applications of McLuhan in our more interactive digital age.

By 1977, I had published two other pieces pertinent to McLuhan's work, in *et cetera* – the Journal of the International Society for General Semantics – which had recently come under Postman's editorship. "Toy, Mirror, and Art" (1977b) presented a full-fledged original theory about the developmental stages of individual media; its relevance to McLuhan and play, work, and art in the digital age is explored in Chapter 11 of this book. That same issue of *et cetera* contained McLuhan's "Laws of the Media" (1977a) along with my Preface to it. (The issue is quite a classic – it also contained Meyrowitz's "The Rise of 'Middle Region' Politics," and its analysis of the erosion of the political hero in the post-JFK television age.) I had come to write such a Preface because Postman had given a draft of McLuhan's article to me, asked for my thoughts, and then for me to commit them to writing. Marshall was of course shown my Preface to his article prior to publication. We met shortly after in New York City, and commenced an all-too-brief period of several years of correspondence and visits in Toronto and New York, including a "Tetrad Conference" I organized at Fairleigh Dickinson University in 1978, featuring Marshall and Eric McLuhan.

There is a discrepancy I have noticed, which almost seems inevitable, whenever I meet in person someone whose creative work I have first come to greatly admire in books and recordings. It is almost as if authors and songwriters pour out their very best in their professional expressions, so that a conversation with them in person cannot possibly measure up to expectations. Marshall McLuhan was the one person I came to know who was an incandescent exception. The walks we took around his home in Wychwood Park, the conversations over dinner, on the phone, in the halls between panels, were every bit as brimming with brilliant insights and wild but somehow sensible connections as his written correspondence and books and articles. Indeed, sometimes a toss-away line in a casual conversation could pack far more punch than McLuhan's written expression, or perhaps clarify it in a way the printed page could not. Throughout this volume, I will thus on occasion quote a nugget or two from our conversations.

The opportunity for continuing such discourse ended with McLuhan's death in 1980. My "McLuhan and Rationality" (1981a), already in press, was published in the *Journal of Communication* the following summer; "McLuhan's Contribution in an Evolutionary Context" (1981b) was published in *Educational Technology* that same year.

"Marshall McLuhan and Computer Conferencing" was not published in the *IEEE Transactions of Professional Communications* until 1986. But it was written in August of 1984, two months after I had logged on and participated in my first online discussions (hosted by the Western Behavioral Sciences Institute) via a Kaypro II CP/M personal computer and a 300-baud modem. The article suggests that McLuhan's oft-criticized, bite-sized, overlapping-little-essay style was in fact a form of online, electronic text, stuck on paper in a media environment which had not yet caught up to the speed and multi-dimensionality of McLuhan's mind. As far as I know, that article was the first by anyone to make a connection between McLuhan and modes of expression in the digital age, and can be considered the seed from which this book arose.

My *Mind at Large: Knowing in the Technological Age* (1988b) was more about technology as an embodiment and vehicle of knowledge than it was about communications media per se; but its intellectual debt to McLuhan's scope and subject, if not his style, was nonetheless enough that I dedicated that book to him. I published letters in *The New York Times* (1981c) and Canada's *Saturday Night* (1988a), pointing out typical misunderstandings of his ideas that continued unabated after his death. My letter to *Saturday Night* also discussed another problem that *Digital McLuhan* by its very existence seeks to address: the incorrect view that McLuhan has had no lasting impact on the world.

In 1990, the *Journal of Communication* published my "McLuhan's Space," a major essay/review of Marchand's biography (1989), Marshall and Eric's *Laws of Media* (1988), McLuhan & Powers' *The Global Village* (1989), and McLuhan's *Letters* (edited and assembled by M. Molinaro, C. McLuhan & W. Toye, 1987), in which I argued for the first time that McLuhan's notion of "acoustic space" was none other than our cyberspace. *WIRED* – aptly listing McLuhan as "Patron Saint" on its masthead – published an updated (but shortened) version of that essay/review in 1993, and McLuhan figured in three of the six articles I published in *WIRED* in the following years (Levinson, 1994a, 1994b, 1995c).

McLuhan's pertinence to the digital age was now well established, if still only barely explored and utilized.

In March 1998, Lance Strate organized a "McLuhan Symposium" at Fordham University. Neil Postman, Joshua Meyrowitz, and James Curtis presented papers, as did I and more than two dozen other scholars who had been stimulated by McLuhan's work as far back as the 1950s. *Digital McLuhan* was mostly drafted then, and the paper I presented – "Way Cool Text through Light Hot Wires" (1998c) – was an adaptation of Chapter 9. A good smattering of the papers touched upon or more fully addressed the relevance of McLuhan to new media, and I strongly recommend the anthology of essays deriving from that conference which Lance Strate and Ed Wachtel are editing for publication, likely in 2000.

In the meantime, *Digital McLuhan* provides an assessment of McLuhan's method (the next chapter) and then 13 of his major insights and what they can tell us about the new world we are well on the way to creating.

Now that McLuhan has been rediscovered – though he was never really lost – I suspect that books about McLuhan, tracts that extend his work, will be appearing with increasing frequency. In such an emerging milieu, I think it especially important that McLuhan's original work be kept close at hand. Interpretations, including this one, inevitably recast the original, and are thus as prone to overlook as bring into better focus an important point.

Accordingly, although *Digital McLuhan* will occasionally quote from an appropriate interpretation, it will rely as much as possible on what McLuhan himself actually wrote or said.

Consider it a continuation of a discussion between McLuhan and me, in which I may have the last word but McLuhan has the upper hand....

2

THE RELUCTANT EXPLICATOR

"I don't explain – I explore," McLuhan explains in "Casting my Perils before Swains," his Preface to Gerald Stearn's *McLuhan: Hot & Cool*, "a critical symposium" on McLuhan's thought published in 1967.

McLuhan had provided what would become a classic example of this method a dozen years earlier, at a seminar he was addressing at Columbia University's Teacher's College in November 1955. He began with a consideration of the alphabet, the x-ray, and Freudian psychoanalysis, and proceeded to the printing press, the telegraph, radio, and television. Louis Forsdale, the seminar leader who had invited McLuhan – hardly known then, outside of those who had read his one book, *The Mechanical Bride* (1951), or who subscribed to the journal *Explorations* that McLuhan edited with his fellow Canadian Edmund Carpenter – took a question from the audience.

It came from Robert K. Merton – the "dean" of American sociology.

"I don't know where to begin," Merton began, nearly purpled with outrage. "Just about everything in your paper requires cross-examination!" He started with the first paragraph, sonorously ticking off all the points in want of further explanation, a William Jennings Bryan making a closing argument to the jury about why the accused should be found guilty of murdering the scholarly procedure.

According to Forsdale, McLuhan took it all in. "Oh, you didn't like those ideas?" he replied with a question, and a twinkle in his eye, "well, then, how about these..."

In today's parlance, we might say McLuhan was just clicking on a link that would move the discussion to another area, whether near or far away.

I I I ■ ■ ■ I I I

But even today, many find such an approach frustrating, maddening. Mark Edmundson (1997), a professor of English, complains that "even McLuhan's most cogent books are wayward and unkempt....They lurch from one subject to another." Yet such frustration – whether Merton's or Edmundson's or ours – stems from a misunderstanding. And precisely because McLuhan had such little inclination to explain, he was in many ways the least best person to address this problem, to elucidate his method.

Merton's misunderstanding is actually easy to understand – and avoid – once we realize that the growth of knowledge, like all developmental processes, consists of more than one stage. Merton was looking for a loaf of bread and a glass of wine, when McLuhan was offering a new kind of grain and a different winepress.

Donald T. Campbell's "evolutionary epistemology" (1974a, 1974b) can help us clarify the difference between McLuhan's offerings and Merton's expectations. The Darwinian model of biological evolution, which continues to serve us well as we move into the new millennium, posits three distinct stages in the evolutionary process: new organic characteristics and organisms arise, independent of the external environment, in the first or "generational" stage (mutation is a part of this process); these characteristics and organisms survive or die based on how well they perform in the world (selection or survival of the fittest – which, by the way, entails cooperation as well as competition as a survival strategy); surviving organisms pass on their genes, and the successful traits they produce, to their offspring (heredity or dissemination). Campbell (and others, before and after, see his 1974a and Cziko & Campbell, 1990) saw that human knowledge evolves in an analogous way: people come up with new ideas (the "generational" stage); these ideas are then exposed to criticism, testing, discussion (the "selection" stage); and those that survive the second stage are published, taught in classrooms, cited in further publications ("disseminated").

So, Merton's problem was that he was expecting McLuhan to partake not only of the first but the second stage – to not only present his ideas, but be party to their criticism, testing, discussion – whereas McLuhan was really interested only in the first.

We can well understand and even sympathize with Merton's frustration, since the academic world – a world still comprised for the most part of books and physical places in which to learn, teach, and study – attempts to operate in all three stages.

But if the academic world, and for that matter, the world at large, concentrates not only on the generation of ideas but their criticism, documentation, and dissemination – and if McLuhan chose to devote his life to the first pursuit, since so much time and attention are already given by others to the second and the third – are we not all better off that he chose that creative course?

The answer would have to be yes – if his ideas were worthwhile.

This book is, in part, both an argument for and testament to that proposition.

But, first, we need to further explicate McLuhan's method – an approach which was both ideally suited to the generation of valuable ideas, yet almost guaranteed to disguise their value to those more interested in criticism and dissemination of knowledge.

ANALOGY AND LOGIC

"It is difficult to make a mistake in logic," McLuhan advises (McLuhan & Nevitt, 1972, p. 8), "once the premises are granted. Psychologists report that madmen are rigorously logical, but their premises are irrelevant. The method of exploration seeks to discover adequate premises."

The method of discovery McLuhan favored – his engine of ideational generation – was analogy, or metaphor. " 'Man's reach must exceed his grasp, or what's a metaphor?' " he quotes with relish an unnamed "wag" 's play on Browning, in "Laws of the Media" (1977a, p. 176). Metaphors indeed increase our reach – that is precisely their contribution, to steer us toward knowledge we do not yet have in hand. Yet they carry baggage which can get in the way of knowledge, when we seek to examine and corroborate it, if we mistake the baggage for the clothing within.

Consider the example of "time flies." It takes the process of time – which no one would mistake for a bird or a plane or Superman – and seeks to illuminate one aspect of it by comparing that aspect to one of the characteristics, probably the most salient, of birds and planes. Thus, "time flies" suggests that time seems to move quickly, just as birds and planes do, certainly from the vantage point of how long it would take us to cover equivalent territory on foot, or even by automobile.

Now, were Merton and Edmundson seated in the audience in which the author of "time flies" first made that comparison, they might well have risen to their feet and objected: Come now, where are the feathers on time, where is the wing on the moment, the propeller on the minute, the engine on the speeding hour? Where is the watch, analog or digital, in the sky?

Of course, no one would ever do that. We have all learned, at least in the culture of the English language (I understand that "time runs" is the equivalent metaphor in Spanish), to decode that metaphor – to take the content of its insight out of the baggage, and not (to use a largely bygone metaphor of the 1960s) get too hung up on it. Or, to spin forth another analogy, we have all learned to enjoy the flesh of that peach for whatever insight into time it can give us – for whatever starting point it might provide in our quest to understand ineffable time – and not frustrate our teeth against the pit, which is what searching for wristwatches in the sky would be for us.

The problem with McLuhan's metaphors, then, is really not that they were metaphors, but metaphors newly minted. Much of the world – including, sadly, the academic – just did not know what to do with them. Worse, and especially in the early days, they did not want to try.

Thus, McLuhan's celebrated observation that "the new electronic interdependence recreates the world in the image of a global village" (1962, p. 43) elicited two kinds of response: one recognized that just as information in a village can be quickly shared by most of its inhabitants, so information transmitted by telephone, radio, and TV could be quickly shared by denizens of our twentieth-century planet. The other response cried out: where on TV are the homes and hearths of the villages we can actually walk into, where on radio is the answer to my personal, specific question that my neighbor in the village might give to me, where on the phone is the face, the handshake, the soft touch of a friend?

Actually, McLuhan – contrary to his reputation for discoursing in "loose, shaggy buffaloes" (that comes again from Edmundson, 1997; see also Sokolov, 1979) – was quite careful with the language in which he chose to convey his analogic insights. Notice that he says, in the above, that electronic media recreate the world in the "image" of a global village, which should give sufficient indication that he is claiming not a complete equation of local village and electronic global community, but an equivalence in some aspect of their informational structures.

Just what that equivalence was, and is, is of course worthy of further research, discourse, and contemplation.

McLuhan was taken seriously at the time of his writings by some people, because they recognized that one or more aspects of one or more of his metaphors hit home, struck a chord of equivalence on some level.

What makes McLuhan even more important as we embark on a new millennium is that the evolution of media since his death in 1980 has sharply increased the match of his metaphors to the reality of our communication.

THE DIGITAL FULFILLMENT, PART I

As Daniel Bell aptly observed (1975, p. 33), "technology is always disruptive" of the status quo. Since metaphors intrinsically exceed the status quo, we can well understand why McLuhan plied and prized them so in his attempts to generate new insights about media: in overshooting the mark, the metaphor gives the mark – and our understanding of it – room to move and grow. In contrast, definitive, fully documented descriptions of a technology, even if they are correct and thus useful in the present, may tell us little about the future. McLuhan was thus fond of quoting Stéphane Mallarmé (writing in 1886): "To define is to kill, To suggest is to create" (McLuhan & Nevitt, 1972, p. 10; see also Karl Popper's dislike of definitions, 1972, p. 328).

The "global village" metaphor, while capturing the immediate, national sharings of information that were already well underway in 1962, missed the mark in many other ways: Telstar, the first telecommunications satellite allowing simultaneous television in Europe and America, was launched in that year, but the regular convening of international television audiences via satellite transmissions was still decades away (see Levinson, 1988b, pp. 117–18, for discussion of early global coverage of the first moon walk, the marriage of Charles and Diana, the funeral of Anwar Sadat, and the Academy Awards; see also Dizard, 1997, pp. 103–4, 107–8, 125). Further, the television audience, even when truly globally seated, was unlike a village in that its members could not converse with one another, unless they happened to be seated in the same physical room – the global village at the time McLuhan coined the metaphor was thus in reality a village of voyeurs, and thus not a village in its important interactive sense at all. But that changed too. And, like the ongoing fulfillment of the global aspect of the global village via CNN and other satellite TV, the fulfillment of the interactive aspect of the global village via the Internet and the World Wide Web was pointed to, suggested, predicted by McLuhan's initiating metaphor.

So, to return to our explication and evaluation of McLuhan's method: Would a Mertonian/Edmundsonian mode of intellectual inquiry, which painstakingly described the worlds of television and radio just as they were in 1962, with the Beatles not even as yet on the *Ed Sullivan Show*, Neil Armstrong not yet upon the moon, the Iron Curtain not yet trampled into the ground by information – would a cross-examined description of that media environment have served us better than McLuhan's metaphor?

Would it serve us better now?

The choice, of course, was never either/or – definition, contra Mallarmé, can live and contribute right alongside of suggestion, and McLuhan was intent not on shattering a mode of discourse but on cultivating an alternative.

His critics – Robert K. Merton (1955, quoted in Marchand, 1989), William Blisset (1958), Dwight MacDonald (1967), Raymond Sokolov (1979), James Morrow (1980), Michael Bliss (1988), to name but a few through the years – saw things differently. And still do (thus, Edmundson, 1997). The very titles of their essays and reviews – "He has looted all culture…to shore up against the ruin of his system" (MacDonald), "Recovering from McLuhan" (Morrow), "False Prophet" (Bliss) – show their contempt, and their unawareness of McLuhan's impact. Bliss's review, written in 1988, after the digital age that would substantiate so much of McLuhan's work was already in dawning view, is especially egregious. Either he or his editors at the Canadian magazine *Saturday Night*, who published his review of McLuhan's *Letters* (Molinaro, C. McLuhan, & Toye, 1987), saw fit to blurb the review, "Once exalted as oracular, Marshall McLuhan's theories now seem laughably inadequate as an intellectual guide to our times." I couldn't resist pointing out, in my Letter to the Editor that *Saturday Night* was good enough to publish in a subsequent issue (Levinson, 1988a, along with similar protestations by George Sanderson, Bruce Powe, and John C. Wilson), that, when it came to writing reviews of books by McLuhan, ignorance apparently was Bliss.

But this special antipathy that book reviewers seem to evince for McLuhan – Sokolov, Bliss, Edmundson, beginning with David Cohn's review of McLuhan's very first book in the *New York Times* in 1951 (Lehmann-Haupt, 1989, is a refreshing exception) – may be no coincidence. Not only was McLuhan's metaphoric method contrary to expectations of traditional scholarship, the very way that he rendered his insights and analogies on the page seemed an affront to the traditional organization of the book.

Indeed, though online text and community at the time of McLuhan's death in 1980 was embryonic at most, the aphoristic bursts of his writing that still so vex his critics seem ideally suited to the Internet and the online milieu. McLuhan, in other words, was writing as if he was contributing to the Web – engaging, in his works of the 1950s through the 1970s, in what by the mid-1980s would become known as "computer conferencing" (see Levinson, 1986).

THE DIGITAL FULFILLMENT, PART II

Writing has readily identifiable patterns of organization. The news story is presented in an "inverted pyramid," with the most important information in the headline, a reprise of the same information with somewhat more detail in the first paragraph, and each successive paragraph containing information of lesser importance; this permits the layout editor to easily shorten any story by lopping off text from the end, in case another story or an advertisement should compete for space. The classic mystery novel begins with a provocative scene – perhaps a dead body – follows with exposition, and eventually an ending in which most or all is revealed. Scholarly texts have their own characteristic patterns as well. They usually begin with an Introduction that sets out a series of themes, episodes, ideas, problems to explore and discuss. These usually have some sort of order – later problems in some sense emerge from the earlier, or at least their discussions are informed by discussions, facts, ideas that were presented earlier in the text. Although the number and length of such units vary considerably in scholarly works, they usually consist of several handfuls of chapters, each more than several pages; each chapter is usually titled with a brief phrase.

A glance at McLuhan's *The Gutenberg Galaxy* (1962) – his second book, and the one that really launched his career as the premier media theorist of the twentieth century – reveals a text which, after its introductory chapter, looks nothing like the traditional scholarly book. Instead of a small number of sequential chapters, *The Gutenberg Galaxy* proceeds with 107 pieces, one to five pages in length, on a series of interlocking, cross-referencing themes. There is some clustering and sequencing of themes – the *Galaxy* moves from the alphabet in the Ancient World through its decline in the

Dark Ages through its re-emergence and differing impact via print in the Renaissance and after – but one can start anywhere in the book, almost in the middle of any slim chapter, and find a set of themes and referents that will serve as ready passport to almost any other part of the book. Each chapter, in other words, contains a blueprint of the entire book, much like the DNA in each of our cells contains a recipe for our entire organism, or the pieces of a holographic plate contain information sufficient to reproduce the entire three-dimensional image recorded on the original, intact plate. So "unchapterlike" are these genomic, holographic sections of *The Gutenberg Galaxy*, that in most cases they bear no title, just a summarizing sentence at their head – e.g., "The increase of visual stress among the Greeks alienated them from the primitive art that the electronic age now re-invents after interiorizing the 'unified field' of electric all-at-onceness" (p. 81).

One need not be a genetic engineer or a holographic photographer to recognize this structure: the number and length of units, the connection of each to the whole, even their untitled headings, bear a remarkable resemblance to much asynchronous online discourse. I noticed this the first month I ever spent online, participating in Andrew Feenberg's "Utopia and Dystopia" course that he taught for the Western Behavioral Sciences Institute in June 1984. At the end of that month, Feenberg and his group of six participants had generated 122 comments ranging from several to more than 100 lines in length; *The Gutenberg Galaxy*, again, has 107 chapter-comments, each about 50 to 150 lines long. (Lines make more sense than pages as measures of online text length, just as they do for the chapters of McLuhan's book, given their very small number of pages; screens or numbers of words or characters also work well as online gauges of text length; Web "pages" have come to mean something quite else, referring to a Web site or to a major subdivision, and encompassing lengths of text ranging from a few lines to whole chapters.)

The WBSI seminar, like all online activity at that time, was conducted in an "ascii" non-hypertext environment. The increasing employment of hypertext links in online fora in the 1990s – links that allow the reader to click on a given word in one text, anywhere on the Web, and be instantly transported to a different text, anywhere else on the Web – has made online text even more like the cross-hatching units of *The Gutenberg Galaxy* published in 1962. Except, of course, that on such an analogy, the whole World Wide Web would be functioning as a single book...

There are other notable differences between McLuhan's writing and online communication. *The Gutenberg Galaxy* ends at 350 pages; online discussions (unlike Feenberg's more structured seminar) can and do go on for years, generating thousands of comments (see, for example, some of the online discussion on the Science Fiction Round Table on Genie, dating back to the early 1990s and still going strong as of this writing). *The Gutenberg Galaxy* was written by one person, McLuhan (so was his next book, *Understanding Media*, 1964, which was a bit more traditional in organization; his subsequent books were all collaborations with one or two people, and were usually even more "electronic" in structure); most online discussions entail at least four or five participants, and often more. Nor can the results of most online discussions by groups of minds compare in originality, scope, and impact with the result of one mind, McLuhan's, as expressed in *The Gutenberg Galaxy*.

These and other differences reflect the fact that *The Gutenberg Galaxy* is, after all, a book, and as such subject to sharp limits in size, as well as the benefits of an author's genius and lifetime of contemplation and research prior to the writing. In contrast, online discussions are usually more casual, both in length and participant intention and research. Such discrepancies, as we saw above, are inevitable in any analogy – the equivalent, in this one, of the feathers missing in time flying.

Nonetheless, even with the discontinuities, McLuhan's mode of writing stands in startling presentiment of online discourse. As far as I know, the early architects of online media (e.g., Hiltz & Turoff, 1978/1993) were in no sense attempting to emulate McLuhan's writing style – indeed, the public and scholarly appreciation of his work was in general decline during this period of the 1970s and 1980s – and clearly McLuhan could not have been emulating an online environment not yet invented in 1962.

So where, then, did this mode come from?

The answer provides a key not only to McLuhan's method of thinking, but what he thought – the content of his work, and, perforce, of this very book.

McLUHAN'S MEDIUM

There is, of course, nothing mystical in McLuhan's prefigurement of online discussion. Both reflect a venerable mode of human discourse – one which, as McLuhan's life's work was at frequent pains to point out, predates the written word. Both reflect that sine qua non of human existence, the mode of communication that no undamaged human lacks anywhere in the world, or lacked (as far as we can tell) anytime in history, namely: spoken conversation.

The connection of online communication to spoken communication is obvious, and has been remarked upon often in recent years (see, e.g., Levinson, 1992, 1995b, 1997b). The ease of entering and changing text on a screen, the nearly instant speed of its transmission – all in contrast to the difficulty and slowness of such applications when conducted on paper – conspired to make online communication a speech-like medium, a hybrid in which our fingers not only do the walking but the talking, from its inception.

McLuhan's interest and grounding in "acoustic space" (see Marchand, 1989, p. 123, and Gordon, 1997, pp. 305–6, for McLuhan's first use of this term in 1954) was less straightforward. Disparate thinkers throughout history have used an aphoristic rather than expository method – ranging from some of the pre-Socratics, not surprising in a pre-literate age, to Nietzsche at the height of the century of letters (see Curtis, 1978, for comparison of Nietzsche and McLuhan). McLuhan's attachment to the acoustic went beyond a chosen method: he saw it as the very basis of human communication (developmentally and historically primal), injured or at very least compromised by the ascendance of written "linear" modes of discourse (worthy of investigation, certainly provocative), and recaptured (though reconfigured) by electronic media (a brilliant insight given the media of his day, prophetic regarding electronic text). Thus, for McLuhan, the method or medium of his presentation was intimately intertwined with its subject. Whether intended or not, the lesson here is that the very form of McLuhan's discourse was an example of what it was attempting to convey.

Might McLuhan have done better to adopt a style more convivial to his academic colleagues, to the world of readers at large?

The common wisdom that McLuhan was his own worst enemy is just that. Insights about acoustic modes of thought presented in single file, rather than herds of "loose, shaggy buffaloes," have fared no better than McLuhan's, and in many cases worse. Vannevar Bush's now famous "As We May Think" (1945) proposed a computer-like device that worked in accordance with the "associative" properties of human thought, in which any given idea could sprout connections to any others; he proposed this "memex" device, an improvement over the straightjacketed file cabinet, in a classically reasoned, sequentially presented, article published in *The Atlantic Monthly*. Merton's reaction to it is not known (at least, not by me), but Bush's article attracted little interest outside of a few visionaries like Ted Nelson (e.g., 1980/1990), until the emergence of hypertext whose need and importance the article predicted made it required reading for every student of the Web in the 1990s.

The advent of "digital culture" is having a similar impact on McLuhan's work as we move into the next millennium. The process has been occurring for nearly a decade, heralded by *WIRED*'s appointment of McLuhan as its "Patron Saint" at its inception in 1993, and sustained by a continuing flow of republications (e.g., the MIT edition of the 1964 *Understanding Media* in 1994) and new works about McLuhan and his work (e.g., Gordon, 1997; see also Strate & Wachtel, forthcoming).

The present book is inevitably part of that rediscovery, but rather than demonstrating how current events make McLuhan relevant, I want to explore – *and* explain – the crucial lessons that McLuhan holds for our digital age. The endpoint is the same – if McLuhan holds crucial lessons for us, then our age makes him relevant – but I want to begin with McLuhan, to minimize the misunderstanding that has dogged so much of his work's reception over the years.

His most famous aphorism – "the medium is the message" – is a case in point.

Not only, as we have seen in this chapter, did the very mode of his presentation embody that principle, but it has been misunderstood as an assertion that the content – in the case of his writing, its subject – is unimportant.

As we will see, McLuhan's view of medium and content was quite the opposite.

3

NET CONTENT

"The medium is the message" is no doubt McLuhan's best-known aphorism. Its fundamental meaning that our use of any communications medium has an impact far greater than the given content of any communications, or what that medium may convey – for example, that the process of watching television has a more significant influence upon our lives than the specific program or content that we watch, or the act of talking on the phone has been more revolutionary in human affairs than most things said on the phone – has been well understood in general, and aptly recognized as the flagstone in McLuhan's path to understanding media. But, unsurprisingly, much of its subtlety and implication has been wildly misinterpreted as a manifesto "against" content, or that what is communicated does not matter at all.

In this case, some of the multiple meanings that were McLuhan's deliberate method – metaphors invite continuing interpretations – were literally spelled out by McLuhan over the years. "The medium is the message" made its first unassuming appearance, as did so much of McLuhan's later headlined thinking, in his 1960 typescript "Report on Project in Understanding New Media" (p. 9) that he wrote for the "National Association of Education Broadcasters pursuant to a contract with the Office of Education, United States Department of Health, Education and Welfare." By 1964, the phrase had become the all-important title of the first chapter of *Understanding Media* – all-important not only because that book would firmly establish McLuhan's reputation as the world's pre-eminent thinker about media, but because, as McLuhan often quipped, no one actually read McLuhan (and to the extent that people read anything, it would likely be not much more than the title of the book's first chapter). So popular were discussions of that chapter title and its meaning, that McLuhan could not resist entitling an entire book with a punning variation – *The Medium is the Massage* by McLuhan and Quentin Fiore in 1967. Two years later, "the medium is the mess-age" appeared in *Counterblast* (McLuhan & Parker, 1969, p. 23), and the "mess-age" became the "mass-age" in *Take Today: The Executive as Dropout* (McLuhan & Nevitt, 1972, p. 63).

Having wrung most of the possibilities out of "message" (though not all – I can certainly see some interesting leads in "the medium is the me sage" or "the medium is ma sage"), McLuhan began playing with the first word – with a result that incidentally punctures his denunciation by critics as arrogant and "untempered by humility" (Bliss, 1988, p. 60). In a car ride with me to the "Tetrad Conference" I organized at Fairleigh Dickinson University in Teaneck, New Jersey in March 1978, McLuhan mentioned that other valuable renditions of his famous phrase might include "the tedium is the message," "the tedium is the mass age," etc. A decade later, John Mortimer (or his editor at *The Sunday Times* in London) saw fit to title his cutting review of McLuhan's *Letters* (Molinaro, C. McLuhan & Toye, 1987) "Tedium is the Message" (Mortimer, 1988), with no attribution to McLuhan or his wit. I suppose it is possible Mortimer or his editor may have come upon this variation all on their own, though McLuhan was not one to confine his word-plays to the privacy of conversations in cars.

In this chapter, we will detail some of the insights "the medium is the message" and its variations hold for our comprehension of media in general and in our digital age. We begin with the crucially important role that McLuhan saw for content.

THE MEDIUM AS CONTENT: WRITING UP ON THE SCREEN

McLuhan's attempt to shift our focus from content to medium derived from his concern that content grabs our attention to the detriment of our understanding and even perception of the medium and all else around it – much as the flood of sunlight on even cloudy days blinds us to the stars that also inhabit our sky, and of which our sun is but a special, particular case. "The 'content' of a medium is like the juicy piece of meat," McLuhan observed in a frequently quoted passage (1964, p. 32), "carried by the burglar to distract the watchdog of the mind." We think and converse about what we read in the newspaper, magazine, or book, what we heard on the radio, what we saw on television, much more often than we do about the fact that we were reading a newspaper rather than listening to the radio or watching television, etc.

Significant exceptions arise when we consider our children – whose media choices we are more prone to monitor – and new media such as the Web, whose general presence in our lives is still unusual enough to be remarked upon, whatever exactly we may be doing with it. Thus, we may well take note of the fact that our child was watching television rather than doing homework, or that we have been spending more or less time on the Web than before. But these are special cases, in which the media come to occupy the prominence in our attention usually accorded content. (Indeed, we might explain McLuhan's initial success as due to the enormous interest everyone continued to have for television as a still-new medium in the early 1960s, when McLuhan published *The Gutenberg Galaxy* and *Understanding Media* and their treatments of television; see also Levinson, 1977b, for more on the public perception of media when they first enter society.)

But what, then, becomes of media when they are eclipsed by content in the more usual state of affairs?

McLuhan provides an answer in two sentences that immediately follow the "watchdog of the mind" line, but are less frequently quoted: "The effect of a medium is made strong and intense just because it is given another medium as 'content'. The content of a movie is a novel…" (1964, p. 32; see also pp. 23–4). In other words, the content of any medium – that "juicy piece of meat" that takes prominence in our perception and distracts us from the deeper impact of the medium at hand – is none other than a prior medium, tamed from its former wild, invisible state, and brought to lie now on our carpet in full view. Not only is the content not unimportant, it may be the best way of examining a medium and its impact – the only drawback being that the medium under such examination will in some sense have arisen earlier than the medium currently in use.

What medium serves as content for the Web, the equivalent of the novel for movies, or movies and the radio serial for TV?

Actually the answer is not one medium, but many media, for the Web has taken as its content the written word in forms ranging from love letters to newspapers, plus telephone, radio ("RealAudio" on the Web), and moving images with sound which can be considered a version of television. But the common denominator in all of these is the written word, as it is and has been with all things having to do with computers – and will likely continue to be until such time, if ever, that the spoken word replaces the written as the vehicle of computer commands. Thus, part of the "message" of the medium of the Internet is all or at least most media that have come before it, with writing ubiquitous in the driver's seat.

Critics of the online age such as Sven Birkerts (e.g., 1994) and Neil Postman (e.g., 1992) are hence not wrong to notice that computers have changed the nature of reading and writing. Surely those activities on paper, in magazine and books, are different than when conducted on screens. But McLuhan's notion of prior media becoming visible, discernible, as the content of new media suggests that one of the differences in the "outing" of writing on the Internet is that the written word is becoming more publicly explicable. And surely the movement of writing into the light, completely out of the monastic shadows from which it had been emerging ever since the introduction of the printing press in Europe, is a good thing – not at all inimical to literacy and its values, as Birkerts et al. claim.

Further, the propelling of writing and other media into new prominence as content on the Internet has another profound and democratizing consequence: unlike our experience with books, newspapers, and magazines, which for all but a tiny fraction of the population has been a one-way engagement of reading not writing, the online experience is two-way, allowing readers to contribute via e-mail, bulletin-board discussion, and all manner of annotation as they navigate the Web. And the growing use of moving images in numerous pages on the Web, including some with accompanying sound, is turning the Web designer into a de facto television producer. The result is an enormous increase and dispersion of image producers, who can render far more images on a public screen with personal computers than can their far fewer counterparts in far more expensive TV and film studios. True, the online image is cartoonish in comparison to TV and film, but it is a product of the world at large, an expression of more and more people with computers, as the written word has been in publicly accessible form since the very advent of the personal computer in the last quarter of the twentieth century.

Indeed, with the exception of the telegraph and the telephone, whose content was always written or spoken by its users, the Internet and its tributaries reverse the trajectory of a handful of messages to a legion of passive users that has typified all technological media since the printing press.

We might say that not only are prior media the content of the Internet, but so too is the human user who, unlike the consumer of other mass media, creates content online with almost every use.

In other words, the user is the content of the Internet – which, it turns out, is much what McLuhan went on to say, in a metaphoric sense, about media in general.

THE USER AS CONTENT ONLINE

"The user is the content," McLuhan often would say to me in the late 1970s (see also McLuhan & Nevitt, 1972, p. 231) – "the CBS eye looks at *you*!" (The CBS "eye" is the logo – the long-running electronic watermark – of the Columbia Broadcasting System television network.) In a 1978 essay in *New York Magazine*, McLuhan further observed that "when you are 'on the telephone' or 'on the air'.... the sender is sent....The disembodied user extends to all those who are recipients of electric information."

Characteristically, we have at least two very distinct though intertwining kinds of relationships described in the above, even in just that last sentence. The person who speaks on the phone or on radio or television creates and becomes disembodied content for the medium – a voice without a face or a body, or, in the case of television, a voice with a face and a body but no substance. In the case of online writing, the process becomes even more streamlined, as users become content in the lines of text they create. This much, I think, is quite clear. But why does McLuhan also refer to the disembodied image on television as the "user," and what is he getting at when he says the TV looks at us, in addition to or rather than vice versa?

This equation of user and content goes back at least as far as the literary criticism of I. A. Richards (1929), who argued that the meaning of a text resided not in its author's intentions but its reader's legitimate (as opposed to idiosyncratic) interpretations. (An idiosyncratic interpretation would be if John thought the novel was marvelous because it took place in London – where he fell in love for the first time last year.) McLuhan took Richards to heart, and moved very reasonably from the user interpreting the text to determining the text to being the text.

But notice that McLuhan's examples of users as content – telephone and television (he also mentions radio) – are all electronic. Telephone, of course, presents a special case, because it is intrinsically interactive, as is online communication. But why distinguish television and radio as media in which the user is "sent"? The answer can only reside in the instantaneity of electronic communication, and the impact it has on the perceiver: whereas books and newspapers bring the world to us, clearly after the fact, radio and TV bring us to the world, to the very scene of the action. The Olympics may indeed be brought into our living room either by newspaper or TV; but to bring the Olympics or a political debate or any unfolding event into our lives by TV is inevitably to also bring us right to the site of the event, to witness its very unfolding. In contrast, aged print on paper, if it brings us to the event at all, does so only after the event is over – after everyone has gone home. (McLuhan further distinguishes the greater participational effects of television in comparison to radio, for other reasons; see Chapters 8 and 9 in this book.)

Thus, McLuhan's notion of the user as content admits to at least a three-part hierarchy in which (a) humans serve as (determine) the content of all media by virtue of our inextricable interpretation of all that comes before us (this can be considered the media equivalent of Kant's view that human knowledge is always the result of our cognitive arrangement of the external world), (b) the human perceiver travels "through" one-way electronic media such as radio and television, and therein becomes their content, and (c) the human conversant literally creates all of the content in older interactive media such as the telephone, and much of the content in the Internet. Further, since the Internet, when it is not trafficking in human conversation (usually in written, now sometimes in spoken, form), presents images in the tradition of television and written words in the tradition of print, it can be seen as a culmination of all that has gone before in the human determination of content – encompassing all three levels.

I have often criticized McLuhan for his media determinism (e.g., Levinson, 1979a), or tendency to cast humans as the "effect" of technology, rather than vice versa (see, for example, his observation that humans are the "sex organs" of technology, 1964, p. 56 – the modern rendition of Samuel Butler's line that the chicken is the egg's way of making another egg, also picked up in the twentieth century by Richard Dawkins and his view of the organism as the gene's way of making more genes). When McLuhan says that the user is "sent," there is no doubt an implication of this subsidiary position of humans in the technological world. But also inherent in McLuhan's schema, and achieving its fullest expression in the Internet, is the human being as an active master of media – not just sent through the media, but calling their shots, literally creating their content, and having unprecedented choice over what that content will be when it has already been created by someone else. My own "anthropotropic" theory of the evolution of media (e.g, Levinson, 1979a, 1997b) – which sees media increasingly selected for their support of "pre-technological," human communication patterns in form and function – is thus much in accord with the resilient humanist element in McLuhan's view of media and content.

The heightening of human choice over media on the Internet – to write, to search, to view, to speak – ties together McLuhan's two complementary notions of content discussed in this and the previous section: the human subscriber of the Internet chooses which prior medium, serving now as content, to use online. But sometimes that choice may be difficult to categorize. If I read a newspaper online, is the newspaper the content...are the words in the online stories...are the ideas expressed in the words...are they all?

The answer suggests that not only do old media become the content of new media, but in so doing retain the older media that served as their content, which in turn retain their even older media as content, going back and back...to the oldest medium of all.

MEDIA WITHIN MEDIA WITHIN MEDIA...THE PRIMACY OF SPEECH AND THE EXCEPTION OF PHOTOGRAPHY

If we were to go back to the first and still most prevalent form of human communication – speech – and ask, what is its content, we would probably say thought, ideas, emotions, and the like. If we were to inquire into the content of those "media" – that is, the medium of thought – we would encounter the contents of the real external world (including the human body and its states), the objects, referents, originals that communication and its re-presentation or representation is all about. In other words, here at the very outset of communication, in the nexus of speech and thought, we come upon a firm distinction between media and content, in which the content – say, a rock we look upon outside of our window – cannot in itself be considered a medium, for it has nothing that we would recognize as content within. (Of course, the rock is comprised of molecules, in turn comprised of atoms, etc. – but this is physical composition, not communication. At the other end of the spectrum, the outside world may contain a radio, which can become the content of our thoughts. But such would be a special case of an element of the outside itself happening to be a medium, and would not count against the general principle that the content of thoughts are "media-less" or "im-mediate.")

As we progress through the history of communication, we find that each new medium takes an older one as its content (as per McLuhan), and that, because of this, speech as the oldest medium has a presence in almost all newer media (see Levinson, 1981d). The phonetic alphabet is a visual representation of spoken sounds. The printing press mass-produces the alphabet in books, newspapers, and magazines. The telegraph sends electronic encodings of written words. The telephone and phonograph and radio obviously convey speech. "Silent" motion photography (which often had musical accompaniment) had visual blurbs of words (along the lines of comic books, and also the "pop-up" videos introduced in the late 1990s); and by the late 1920s, it had quite literally begun to talk. Motion pictures (along with elements of radio, including serials, news, and the network structure) became the content of television. And all of the above are rapidly becoming the content of the Internet, the medium of media. Speech is thus ever with us, not only because it is the form of communication that most people use most of the time directly, but because it is present as the content of all subsequent current media – with one exception.

That exception would be still photography, which takes as its content not the spoken word, but the outside world itself – in other words, what speech and thought are usually about. The truism that a picture is worth a thousand words recognizes this privileged position: photography can be worth much more than words precisely because it does what words do, without words, in a very different way. Indeed, still photography partakes of a non-verbal iconic mode of communication that goes back to the cave paintings in Chauvet, Lascaux, and Altamira, and weaves its way through the hieroglyphics of Ancient Egypt and the ideographic writing systems still in use today in the Far East. (Motion photography does not, because its content is usually not the world as it is but a narrative – or, as McLuhan pointed out, a novel.)

But significantly, even the analogic photograph, whose connection to its subject is literal not symbolic – the image on the photograph comes from light that literally bounces off the external world, and thus copies rather than describes it – has been situated and framed in the verbal almost since its inception. A picture may be worth a thousand words, but when printed it almost always is accompanied by a written caption.

And these iconic islands in a sea of symbolic words are now undergoing a full digital reconfiguration: once the photograph is converted to a digital format, it is as amenable to manipulation, as divorced from the reality it purports to represent, as the words which appear on the same screen. On that score, the Internet's co-option of photography – the rendering of the formerly analogic image as its content – is at least as profound as the Internet's promotion of written communication.

It is a victory for the digital foreseen by McLuhan, although he did not refer to it as digital. What he did see was the profound tension between what he called "acoustic space," along with its suggestive unseen possibilities, and "visual space," which, with its fixed point of view, was both more precise and limited – in the way that a written description of a meadow can be far more expansive (or "acoustic") than a pinpoint photograph.

As will be the case fairly often throughout this book, we will see that some of McLuhan's categorizations work better with examples of media different from the ones he used, or with examples even shifted from one category or field to another. This is only to be expected, since the fundamental thesis of this book is that the digital age is both well explained by McLuhan and helps bring McLuhan's ideas into sharper focus – i.e., McLuhan makes more sense when reconfigured for the digital age. Thus, McLuhan (1962) characterized the alphabet as the main architect of visual (not acoustic) space – because it is visual – and competes with, draws power from, the spoken word. I would make the cut differently: since the alphabet literally looks nothing like the world and events it describes, nothing like its content, it is actually acoustic – or, as I say in *The Soft Edge* (1997b), the first digital medium. Or perhaps not the first, if we are willing to consider DNA a medium of sorts: for it too bears no resemblance whatsoever to the myriad of biological structures and entities it instructs proteins into becoming.

What McLuhan did see with crystal clarity is that the acoustic (digital) was the far more powerful and encompassing of the two modes of representation, and via the electronic revolution was steadily gaining ground over literal, visual communication in the twentieth century. He called the world it was fetching into being, and retrieving from pre-literate times, acoustic space.

As we will see in the next chapter, it is as much literate as pre-literate, and in today's parlance is known as cyberspace.

4

THE SONG OF THE ALPHABET IN CYBERSPACE

According to Philip Marchand (1989, p. 123), McLuhan first broached the topic of acoustic space in a paper read by Carl Williams at McLuhan and Edmund Carpenter's 1954 seminars in Toronto. (Marchand says the paper had "the stamp of McLuhan all over it," and implies that its author was McLuhan not Williams; Gordon, 1997, pp. 305–6, says Williams conveyed the idea from E. A. Bott, a professor of psychology at the University of Toronto, who first developed it.) In any case, the paper was published the following year in the journal *Explorations* under a Carpenter and McLuhan byline, and reprinted in the 1960 anthology *Explorations in Communication* – still one of the most cogent presentations of many of McLuhan's major insights.

The paper makes a startling point: what we consider normal or natural visual space is actually a technological artifact – a result of perceptual habits created by reading and writing with a phonetic alphabet. Or as McLuhan put it, much later, in two books posthumously published, "when the consonant was invented as a meaningless abstraction, vision detached itself from the other senses and visual space began to form" (M. McLuhan & E. McLuhan, 1988, p. 13), and "visual space is a side effect of the uniform, continuous, and fragmented character of the phonetic alphabet, originated by the Phoenicians and enlarged by the Greeks" (McLuhan &

Powers, 1989, p. 35). Thus, McLuhan, whether with colleagues (Carpenter, Powers, Bott and Williams), son (Eric), or on his own, is consistent in his view that what we take for granted in the shapes and organization of our external visual world is actually a consequence of the technological lenses through which many of us for the past 2,500 years of Western history have been inclined to regard the world – specifically, the prism of the linear, connected alphabet. In the parlance of the previous chapter, we might say that visual space is a kind of content of an artificial medium. Or, as Patrick Heelan (1983) notes with a remarkably apt though apparently coincidental choice of terms, our commonsense Euclidean perceptions of the world – what we feel to be intuitive notions of space and time – are in fact the result of our technologically "carpentered" world, or the way that we and our ancestors have constructed dwellings since we emerged from caves (Heelan mentions neither Carpenter nor McLuhan in his book).

Certainly, if we agree with Kant that our perception of the world is as much a result of our perceptual/cognitive structures ("categories," in Kant's usage) as the stuff of the external world itself – that a color-blind person sees the same world quite differently from a person with full-color vision, who in turn may well be "color-blind" to other aspects of external reality – then we can see the merit of McLuhan's, and Heelan's, general positions. But which of these positions is actually closest to the truth? Do we see the world in right angles because of what Kant would call an innate and we might today refer to as a genetically provided tendency to view the world that way – in much the same way that color-blindness is biologically determined? Is that also why we "carpenter" the world in lines and angles – is that a reflection of some innate Euclidean tendency? Or is this rather a case of form following practical function, in which lines and angles work better, are easier to construct, than softer, blurrier, non-Euclidean forms? And does the alphabet, as McLuhan suggests, drive this Euclidean process – or is it the other way around, with the alphabet and its success more of a consequence than a cause of the rectangled world?

Or is the alphabet perhaps driving in a different, non-Euclidean direction altogether?

McLuhan's notion of "acoustic" space makes his position on visual space and its relation to the alphabet very clear: acoustic space comes before alphabet. It is the world viewed through pre-literate eyes, a world of no boundaries in which information emerges not from fixed positions but anywhere and everywhere. It is the world of music, myth, total immersion.

And it is thus, on McLuhan's view, also the world that comes to us after the alphabet, in the form mostly of television, which is also musical, mythic, immersive, and, unlike the book and the newspaper, lacks perspective or distance from its subjects.

But what, then, of cyberspace – the world that is both alphabetic, yet comes after the world of television, and indeed is beginning to subsume it?

In this chapter, we stand McLuhan on his head, and argue that his acoustic space is now most found in the online alphabetic milieu of cyberspace.

THE ACOUSTIC WORLD, HEARD OFFLINE

To evaluate my contention that cyberspace is acoustic space – that the alphabet online has as much in common with music in the dark as with print on paper – we need to consider (a) the nature of acoustic space prior to the alphabet, and (b) the nature of the alphabet prior to its emigration online. In this section, we look at pre-alphabetic acoustic space.

The characteristics of acoustic space derive, unsurprisingly, from the qualities of hearing, in contrast to seeing, tasting, touching, and other modes of perceiving the world. Taste and touch, for example, require direct physical contact with their objects of perception. Such "amoebic" modes of perception (see Campbell, 1974a, for more) are usually highly truthful – we rarely suffer from tactile illusions – but somewhat dangerous, since we and the amoeba might well die of physical contact with a noxious substance. Seeing, at the other end of this continuum, affords us the safety of distance and detachment, but with a price-tag of much greater likelihood of error, in part because we are prone to focus on one aspect of the environment to the exclusion of all else, in part because we are more likely to misunderstand what we are seeing than what we are touching (common optical illusions are the most extreme example of this). Hearing, on this analysis, is the happy medium, in that it gives us some of the distance and safety of sight, without as much loss of background and immersion.

Our very language pinpoints some of these differences in perception: we may look at something or see it, listen to it or hear it, but only touch, taste, or smell it – not touch "at" or taste "to" it. By that criterion, seeing and hearing are on the mediated side of a divide, in which an "at" or a "to" can intrude between the perceiver and the object. Touching, tasting, and smelling brook no such mediation – nothing comes between the toucher and the touched. But whereas seeing seems to revel in its distance – to see or look at something is intrinsically to map where it is out there in relation to us – hearing often behaves as if it is a form of unmediated touching. We swim through a world of sounds, which approach us from all sectors of the environment, whether our ears are focused in the direction of the sound or not. And though we might well make inferences about our distance from the source of a sound (as we might also about the source of a smell, which is more like seeing and hearing than touching and tasting on this account, in that we need not be in physical contact with the origin of the smell), our more immediate sense of sounds is probably that they are loud or soft, equally close to us, with the imputation of distance for some kinds of soft sound a secondary assessment that we bring to the perception.

A related characteristic of sound – and one that distinguishes it from the objects of sight, touch, taste, and smell – is that it seems to be with us on all occasions, emanating from every environment. The world grows dark every night (and literally punctuates sight) but never really silent; we close our eyes (another kind of visual punctuation) but not our ears. Touch and taste are even more specific and delineated than vision in their performance: we touch or taste only that which is upon our skin or tongue. With no proximate stimulus, those senses are unengaged. (A breeze upon the skin, or an uncomfortable item of clothing, probably provide the most continuous experiences of touch, but even those are far more keyed to a specific environment, with a beginning and an end, than the continuing serenade of one sound or another in our world.) Smell is more diffuse than touch or taste, but it plays such a minor role in the human sensorium that most of the time we discern no significant fragrance or odor at all.

We scout the world, then, through vision and hearing (and to a much lesser extent, via smell); we engage the world via touch and taste. As scouts, both vision and hearing give us first reports on aspects of the world with which we are not yet engaged. But they scout in radically different ways. Whereas vision gives us precise, on-the-spot details about premises on which we bestow our sight, hearing keeps us in contact with the world twenty-four hours per day, whether we choose to bestow our ears upon that world or not. Sunlight streaming through the window is thus usually insufficient to rouse us from a deep sleep – we instead require a mercilessly ringing alarm clock, which plays to that part of our sensorium never on break, respite, or holiday. Were it not for hearing, we likely would not have survived too many dark nights as a species. We owe our survival to this constant monitor, this indefatigable eavesdropper prepared to convey information at any and all times.

Which brings us to the evolution of the alphabet into an analogous kind of constant conveyance.

SCRIPT, PRINT, CYBERSPACE

The alphabet has had a hard time living down its origin as a solitary medium, texts of which were usually readable by no more than one set of eyes at any one time, and capable of being copied only by dint of someone literally transcribing page per page. Even the prehistoric picture, pictograph, and hieroglyphic on the wall or monument were usually viewable by groups, however limited, prior to the alphabet. In that important sense, then – the sense of simultaneity, or being present in more than one place at the same time – the alphabet at first was a partial step backward. This, of course, is only to be expected, as evolution whether of biological or technological species and characteristics is always a tradeoff rather than a clear, one-hundred-percent jump forward, with improvements net and on-balance not complete (see Levinson, 1979a, 1988b, 1997b, for more; human intelligence in general may be an exception – a complete advantage over its predecessors in evolution – but individual technologies are more likely to be improvements only insofar as advantages outweigh disadvantages).

The alphabet's segregating tendencies, however, were reversed to some extent by the printing press. Stradanus' (1523–1605) caption on his engraving of a bustling early print shop captures its acoustic impact: "Just as one voice can be heard by a multitude of ears, so single writings cover a thousand sheets" (reprinted in Agassi, 1968, p. 26). The increasing result of this publication was many more than one copy of alphabetic texts available to many more than one person – to "publics" rather than disparate people – especially in the case of high-circulation newspapers and best-selling books found on most newsstands and in many bookstores and libraries.

On the other hand, such dissemination via print was never as easy as the voice "heard by a multitude of ears," nor as far-reaching as the voices and faces conveyed by the electronic media of radio and television. Further, the vast majority of books published are not and never were best-sellers – meaning that they are not available in most outlets, certainly not bookstores. Thus, although McLuhan can be faulted for lumping mass-produced texts along with handwritten manuscripts as alphabetic-visual culture, and overlooking the incipient electronic-acoustic properties of print, the sin is certainly not cardinal.

What of course is acoustic about television, which McLuhan did hail as such, is that the same images can be seen on television screens anywhere in a given country – and, since the advent of cable such as CNN, anywhere in the world. Radio shares this figurative acoustic effect, as well as being literally acoustic. Once the alphabet began to appear on screens – screens which, by virtue of their connection to the Internet, can display the same words (say, a given page on the Web) even more effortlessly and efficiently on a worldwide basis than the news on international cable TV screens – the alphabet became both the content and proximate conduit of an acoustic space far more intimate and immediately distant in its reach than print.

As we saw in the previous chapter, the content of a medium – or of an environment such as acoustic space – is not insignificant. Indeed, it operates as an intra-medium, or environment within environment, and helps define the rules of our engagement with the overall medium. Thus, we have been considering at least four distinct kinds of acoustic space in the above discussion – unmediated hearing, radio, television, and cyberspace (with print perhaps comprising a partial kind of acoustic space, contra McLuhan) – each with its own variations on the acoustic theme.

Indeed, the rendition of acoustic space in alphabetic cyberspace departs in several crucial ways from those older media.

THE ACOUSTIC WORLD, SEEN ONLINE

The alphabet is not the only medium available on the Internet, but all online media share one of the alphabet's original and still key characteristics: the volition it facilitates in its writers. Just as almost any individual – other than a very young child – with reed, quill, or pen in hand can fill the blank page with whatever letters are desired (in contrast to paintings and even hieroglyphics, which require special talent and/or years of training to render well), so too does almost any individual in front of a blank online screen have enormous power over what transpires upon it.

We can see this clearly in the sharp increase in control that the listener has over radio when it is online (in "RealAudio" on the Internet) in contrast to radio broadcasts offline. Once archived in an online file, the radio program in "RealAudio" is available at all times to anyone anywhere in the world with a computer and a modem – unlike the traditional radio broadcast available only at specified times and for limited distances (unless it is recorded by the listener, which requires an additional technology and can only preserve a tiny number of broadcasts in comparison to those available on the Internet). Thus, the acoustic space of offline radio undergoes an "alphabetic" transformation online, in which simultaneity at the time of the replay is diminished (reduction of an acoustic quality) but user control is augmented (increase of a different acoustic quality). But since the original radio program online – before it is archived – can enjoy the same simultaneity as the classic, offline radio broadcast, "RealAudio" represents a real or net increase in acoustic space.

As the governing medium of cyberspace, the alphabet imparts its most intrinsic qualities, such as increased user control. Other qualities – such as the oft-noted non-interactivity of writing in contrast to speaking – are left behind with the alphabet on paper, and indeed turn out to be not qualities essential to the alphabet or writing at all, but rather the requirements of paper and the alphabet's other earlier physical partners. The alphabet, in other words, finds its most purely or characteristically alphabetic expression in the acoustic space it creates online.

Indeed, the easy interactivity of online alphabetic communication – instant from anywhere in the world in synchronous live chats, available anytime night or day in asynchronous discussions (see Levinson, 1995b, for more on this distinction) – retrieves and augments a fundamental acoustic dynamic missing in every medium since unmediated hearing, save the telephone (see Chapter 15 in the present book for McLuhan's notions of retrieval and enhancement/augmentation in media). The eyes that became ears in the print shop of Stradanus nonetheless lacked mouths: readers of printed books got the same unvarying answers to their questions – namely, the answers already printed – as the readers of handwritten manuscripts lamented by Socrates in Plato's *Phaedrus*. Similarly, questions posed to one-way radio and television broadcasts fall on deaf ears (unless the broadcast or cable show is a "phone-in" – designed to take calls from listeners and viewers via telephone). Thus, radio and television and products of the printing press are cases of "closed" acoustic space (letters to the editor being a minor exception), or on that criterion not acoustic space at all, if we view openness and interactivity as one of its defining features. In contrast, online communication approximates at least part of the openness of the in-person conversation.

Of course, there is a crucial element of online alphabetic communication which is uncompromisingly visual: we absorb its information through our eyes. And that means we must give it, for the most part, our undivided attention. We cannot partake of cyberspace with eyes closed, or when driving, or rushing around with a glass of orange juice and a bagel in the morning, as we certainly can radio (its signal advantage over all other media) and even television (except when driving). Indeed, the little that television asks of us – we can doze off in front of it, make love in front of it, and still receive much of its communication – is one of its great benefits, as I pointed out nearly two decades ago (see Levinson, 1980; see also McGrath, 1997, for a similar defense of TV). Not every informational engagement has to command our full attention; as living organisms, we also require a significant amount of time off (sleep is the most obvious example, but there are of course numerous states between sleep and full alertness). Media which accompany us at such times thus have their value too. This low-key, take-it-or-leave-it quality certainly feels acoustic, though it is more intrinsic to radio and television than in-person conversations, which do require a minimal level of attention from both parties.

Looking to the future, we could imagine a cyberspace with less or no alphabet – an online communication system comprised in part or entirely of speech. Such a system would no doubt be more acoustic, or acoustic on more levels, than current alphabetic cyberspace. But envisioning it also calls into question the very status of the alphabet: is it, was it, just a convenience, the very best we could do for some kinds of communication in a world lacking more efficient media? Or was it, or has it become, something more?

A CYBERSPACE *SANS* ALPHABET?

McLuhan had a keen sense of history, and, perforce, of the future. In recognizing the extent to which current media retrieve communication systems of the past – as in the global village, or, for that matter, the notion of television as acoustic space – McLuhan invited us to think about the ways that current media may someday be retrieved in the media of the future.

But McLuhan provides no underlying theory of why this should be – no explanation of the overall evolution of media, to what ends the mechanisms he describes may be working, what deeper forces may administer their operation. Thus, other than offering us the possibility that the alphabet may someday be retrieved – if ever it falls completely out of fashion – McLuhan provides little advice as to its ultimate fate. Similarly, we can only guess at the future of cyberspace via McLuhan, or at most choose its forthcoming constituents from a platter of past media.

For the past two decades, beginning with my "Human Replay: A Theory of the Evolution of Media" (1979a), I have been developing a general theory which might help us in the difficult task of predicting the future of communications. The crux of the theory is this: media evolve in a Darwinian manner, with human beings acting not only as their inventors (obviously) but their selectors (i.e., the selecting environment, in Darwinian terms). We make our selections on the basis of two criteria: (a) we want media to extend our communications beyond the biological boundaries of naked seeing and hearing (this only restates McLuhan's view of media as "extensions" across time and space – a view which he in turn had constructively adopted and expanded from Harold Innis) (see McLuhan, 1964; Innis, 1950, 1951); (b) we want media to recapture elements of that biological communication which early artificial extensions may have lost – we want, in other words, our hearth of natural communication even as we exceed it in our extensions.

This second component, although it encompasses McLuhan's notion of retrieval, goes well beyond it in specifying what elements of communication are most likely to be retrieved. Telephone replaces telegraph, under our evolutionary pressure for this two-way communication system to retrieve the lost element of voice. Color replaces black-and-white on all fronts of communication except the deliberate artistic statement, because we yearn to see the colors of the natural world in our technological reproductions of it; similarly, talking motion pictures replace silent (actually, speechless) motion pictures. Radio, however, is not eradicated by television, because it turns out that hearing without seeing – the eavesdropping function mentioned above – is a significant component of our natural (pre-technological) communication environment, and perfectly played to by radio. In all cases, those media that stand and continue to thrive over time do so because they replicate, correspond to, accommodate, retrieve, an important facet or mode of unmediated, biological communication.

What, then, of the alphabet, on this "anthropotropic" ("anthropo"= human; "tropic"=towards) theory of media evolution? Clearly, there is no alphabet in the pre-technological, naked human world. If there were, we would not need to devote at least the first six years of schooling to teaching our children to read and write.

In contrast, speech requires no such schooling. It can be improved via tutelage, certainly; but as Noam Chomsky conclusively demonstrated nearly a half century ago, children learn how to speak simply by hearing speech in an off-hand, unprogrammed way. Or better, we come hardwired with a program for speech, which is easily elicited by an appropriate environment – namely, hearing other people speak. Speech, then, is as quintessentially human as what our eyes do for our vision and our ears for what we hear.

But speech entails a uniquely human process which seeing and hearing do not. Whereas we are prone to see and hear the world in some sense as it is – as it presents itself to us, in the same differing but overlapping ways it presents itself to all living organisms – speech is prone to present the world to us as it is not. We can and do speak of things not physically present – of a past, future, or generality such as justice – much more easily than we can see or hear them. Indeed, absent speech, we likely could not see or hear much more than the immediately physically present world at all, other than the occasional recollected image or melody that presents itself unbidden to the mind's eye and ear. (Kant is of course right that our processes of vision and hearing mold that immediate physical world – and thus the world presented to us is the world that our senses present to us, and shape to some extent in the presenting. But this molding, by and large, does not in itself give us a sense of past, future, or generality.)

That generalizing quality of speech – the capacity that gives us a sense of things not present, the world as it is not – is abstraction. So fundamental is it to our language, our humanity, that we can hardly imagine the two without it. Speech minus abstraction does exist in screams, shrieks, cries of joy and pain – these indeed tell us of immediate, proximate states – but these could only be considered language in the most metaphoric or "proto" of meanings.

My "anthropotropic" theory, then, would predict that preservation, retrieval, and enhancement of our capacity for abstraction would be a crucial selecting pressure in the evolution of media. But, with the exceptions of the alphabet, the printing press, and the telegraph, most media prior to the Internet have not done this. Instead, painting, photography, telephone, the phonograph, motion pictures, radio, and television have each in their own ways brought us more of the world as we see and hear it with our eyes and ears. Why, in view of the centrality of abstraction in our lives, has so much of media evolution been iconic?

The answer, I would argue, is that the alphabet conveys abstraction so effectively that we lack the impetus to improve upon it in other media. Print, of course, is but the alphabet writ large; as is the telegraph, in another sense. The alphabet is more abstract than even speech, its progenitor, which conveys non-abstract emotional tone in the quality of the voice speaking the words. (Hence, McLuhan's accurate depiction of the alphabet as a "flattening" agent; see McLuhan, 1962). And the alphabet is more abstract at its actual point of usage than current digital media – which, although highly abstract on the programming level (the binary codes that can represent sounds, images, letters), often operates iconically on the usage level (as when we see pictures and hear sounds on the Internet).

Given this uniquely high degree of abstraction of the alphabet, together with the centrality of abstraction to human thought and life, a reasonable prediction based on a Darwinian evolution of media towards increasing consonance with human communication would be that the alphabet's place as the conductor of acoustic cyberspace is quite secure.

Of course, we and our societies evolve along with our media. In the next chapter, we look at what happens to our bodies – their role in our lives and society – when our representations, our image and acoustic proxies, abstract and iconic, do our bidding in cyberspace. We look at what happens to us in electronic media when "the sender is sent" (McLuhan, 1978).

5

ONLINE ANGELS

Moving great distances instantly – or at speed of light, which is the equivalent of instant for distances on Earth – has been a mainspring of science fiction since the second half of the twentieth century. Alfred Bester had Gully Foyle do it via "jaunting" in *The Stars My Destination* (1957), and of course Scotty and Geordi and other engineers in the *Star Trek* universe have been "beaming" people around for decades.

The impetus for such teleportation predates science fiction. Angels, spirits, demons, and all manner of other-worldly entities traverse the world instantly throughout history. And why not? To imagine conversing with someone, being in that person's physical presence, regardless of how far away he or she may be, is as natural as imagination itself: in the intensely compressed information system of the human brain, the time and distance of its constructs are often irrelevant.

In the real, external world, however, time and distance count. Thus, although we have devised means of moving our bodies ever faster – well surpassing speed of sound in the twentieth century – we have as yet invented no way of speeding our bodies the slightest bit forward or backward in time. And the fastest means of physical transport are nothing close to instant – that speed being a feature of communication, not transport, since the telegraph in the nineteenth century. (See Levinson, 1982, 1994b, for a brief discussion of why time travel may in principle always be impossible, but faster-than-light travel may not.)

Indeed, although communication and transportation were once intrinsically overlapping acts – talking to someone required walking into that person's immediate physical environment, although one could of course talk about things not physically at hand – communication ever since writing and indeed cave painting has distinguished itself by accomplishments unattainable through transport. Thus, Freud (1930, p. 38) hears in writing "the voice of an absent person," and André Bazin (1967, p. 14) sees in the photograph an image rescued from "its proper corruption" in time (the same could be said of the images of bison and other animals upon Cro-Magnon cave walls, although the rescue was presumably more subjective, i.e., the image was more a function of the painter – as it is with painting today – than the photograph is of the photographer). And just around the time of the photograph's debut, the telegraph made its speed-of-light entrance. Indeed, Samuel Morse worked on both inventions, even though he brought only the telegraph to consummation (L. J. M. Daguerre did the honors for the photograph).

The content of the telegraph was triply abstract – electronic signals which represented letters of the alphabet (as in Morse Code) which themselves represented spoken words, which in turn represented reality. And each of these representations was arbitrary, unlike the direct literal connection between the initial photographic image and its object. Thus, we can contemplate only with difficulty McLuhan's notion of the "sender being sent" in the telegraph, since what was sent had such little physical resemblance to the sender.

But beginning with the telephone, the sent sender becomes increasingly embodied. The telephone resupplies the voice at instant distance, as does radio, in a broadcast rather than an interpersonal mode. The broadcast is significant, in that it conveys representations not only instantly, but simultaneously to many people.

"Nixon on TV is everywhere at once," Edmund Carpenter wrote (1972/1973, p. 3) at the height of Nixon's popularity and power in 1972. "That is the Neo-Platonic definition of God." Television, obviously, restores the missing image.

In the 1990s, the Internet sent mainly the written words of its users, in a mode both mass (broadcast) and interpersonal – one-to-one – at the same time. As we move into the twenty-first century, the Internet increasingly integrates voices and images into its instant, simultaneous mix.

But none of these sendings approach the full embodiment of transportation, nor do they bestow upon their passengers the power over events that they might have hoped to enjoy. Nixon the virtual God fell far short of Augustine's, and indeed the very TV that made him a god helped bring him down in its unblinking coverage of his transparent denials of the Watergate cover-up. The medium that made him an angel, metaphysically speaking, conspired to make him a fallen angel, a Lucifer of our time (Nixon's resemblance to Mephistopheles had already been remarked upon after his appearances in the Presidential debates with John F. Kennedy in 1960).

As we encounter the new millennium, we are left, then, with instant communication of embodiments reflective enough of the real thing to be taken seriously, but still incomplete, and, like all media, not totally in anyone's control.

What happens to bodies – online, on television, on radio and the phone – absent their physicality? What impact does this have on us, the physical living beings represented or "sent" in those virtual surrogates?

This is the subject of McLuhan's "discarnate man," and this chapter.

VIRTUAL VIRTUE AND SEX DISCARNATE

McLuhan's points of departure for the devil discarnate – I prefer that term to discarnate man, because it coincides with the godlike attribute of being everywhere at once (and is gender neutral) – are the corporeal qualities this etherealization leaves behind. The human being on the air, on the phone (and now online) "has a very weak awareness of private identity," McLuhan explains (1978), "and has been relieved of all commitments to law and morals." Carpenter (1972/1973, p. 3) makes a similar observation, indirectly, when he qualifies his view that "electricity has made angels of us all." He adds, "not angels in the Sunday school sense of being good or having wings, but spirit freed from flesh, capable of instant transportation anywhere." The qualification of not in the "sense of being good" is important, because in the traditional Platonic/Christian/Buddhist et al. view, flesh is weak, degrading, mortifying, in contrast to the purity of the soaring spirit. But not when the spirit soars on air or online.

In what sense do the spirits of television leave behind morality? McLuhan (1978) saw loss of personal identity and urban violence as consequences of this amoral media state, and suggested that "all the fantasy violence of TV is a reminder that the violence of the real world is motivated by people questing for lost identity." Although this is an innovative presentation of a widespread assumption that violence in the media, on television in particular, is responsible for violence in the real world (whether via suggestion of violence to otherwise innocent minds, or frustration of viewers by advertising unaffordable products, or some combination), it is nonetheless unsupported and even contradicted by evidence: studies fail to distinguish between TV as a cause versus a reflection of violence, societies such as Canada with more or less the same TV programming as the United States have had much lower levels of violence (closer to those low levels in England) than the United States, and so forth (see Levinson, 1994a, for more). Indeed, the plummeting of murder rates in late 1990s New York City to those of the 1960s (Butterfield, 1997), with violence still being featured nightly on popular television programs such as *Homicide*, *NYPD Blue*, and *Law and Order*, pulls down another girder in the supposed bridge from violence on television to violence in the real world.

But, as we have seen and will continue to see throughout this book, what works as a shaky metaphor for McLuhan regarding television suddenly finds vivid fulfillment in the online digital world – where personal identity can indeed be easily jettisoned, albeit with no necessary increase in violence.

The plaything of the online spirit, freed from corporeality, is not thanatos but libido: sexuality. The French, appropriately enough, first noticed this en masse with their "Minitel" users in the mid-1980s. Developed initially as a quite sensible replacement of cumbersome paper phone directories which were distributed yearly, the Minitel's "rose" message boards quickly became a thicket of sexual encounter – conducted entirely via text. People met late at night online in what we today call "live chat rooms," and described in scintillating detail what they were imagining doing to each other's bodies (see Levinson, 1985, 1992, and Rheingold, 1993, for more).

But, of course, there are no bodies online, and in the 1980s there were not even images of bodies online. (McLuhan, if he were writing in the 1980s, might well have noted, "online, everybody is a nobody.") The bodies were at home, presumably alone – clothed or not, sometimes masturbating – but in a dimension utterly other from the online locus of the sexual encounter. In some cases, the parties might have already known each other, and thus what they looked like. But the purest instances of this genre occurred between people who had never seen each other in the flesh, or in camera. The opportunities for deception were of course enormous – people dissembling or outrightly lying about their looks, age, even gender

– but as long as there was no corporeal follow-through, as long as the parties kept their love-making virtual with no attempts to confirm in flesh offline – the deceptions were self-contained (see van Gelder, 1985, and Levinson, 1997b, p. 156, for examples that did spill over into the analog world, with commensurate problems).

The possibility of virtual interpersonal misperception – whether intended by one or both parties or otherwise – predates online communication, and harkens back to the telephone, telegraph, and any written communication in which the correspondents are physically unknown to one another. I recall well a good friend when I was 14, who had a "pen-pal" overseas. At someone's suggestion, they exchanged photos; my friend was displeased with what he saw, and never wrote to her, his pen-pal, again. Ah, callow youth! But two years later, another friend of mine – let's call him George – called me on the phone one day with the exciting news that he had reached a wrong number a few minutes earlier, and found a girl there with a wonderful, sexy voice. They proceeded to talk on the phone for weeks and weeks, sometimes for hours at a time. Eventually – at my helpful suggestion – George decided he had to see this girl. He was falling in love with her, and could not go on any longer without knowing what she looked like. At first, she demurred. But, eventually, she agreed. (I came along as moral support, and she also had a friend by her side.) Almost needless to say, George was sorely disappointed. It took him months to get over this. (Don't ask me why I twice had friends with experiences such as this – perhaps it's part of what led me to become a media theorist.)

The reason George's disappointment was almost a foregone conclusion is that his imagination had supplied the perfect image for the voice he had all but fallen in love with. In this sense, the discarnate experience is indeed quintessentially Platonic, in the original meaning of the term: in a love affair of just written words (or words spoken without faces), our minds provide the missing parts, and make them perfect, like all denizens of Plato's ideal realm. The only defense is resisting the fuller, in-person encounter, and therein preserving the Platonic ideal from contradiction in the flesh.

Nor is this necessarily immoral. As AIDS became more prominent in the late 1980s and 1990s, the deceptions of phone-sex and online liaisons became attractively safe. Computers are prone to "viruses" too – but these at most can cost you your computer's life, not yours. Meanwhile, the growing use of sounds and images online may be removing some of the potential for deception from virtual reality, though a clever programmer could no doubt rig his or her machine to present a false voice or image.

Thus, in as much as sex online entails the risk of neither pregnancy nor disease, it is indeed more angelic "in the Sunday school sense" than sex the old fashioned way. Moreover, as long as online angels take care not to jeopardize their status by falling offline, or coming down to Earth to confirm or extend their relationship in the real, palpable world, their safety and lightness of being is assured. The relieving of "all commitments" online has a silver lining.

But there is more to human life than sex – in and out of Sunday school, on and offline. What impact does moving beyond our bodies via telecommunication have on other aspects of our existence – for example, on our view of where we stand in the universe?

GOD'S-EYE VIEWS

It was Sigmund Freud (1930, pp. 38–9) who observed that when we don our technologies, we "become a kind of prosthetic God." Freud had in mind much more than just electronic media, noting that boats, aircraft, eyeglasses, and photographs – as well as telephones and writing – all conspire to make us "truly magnificent," if not particularly happy or well-adjusted to this God-like stature.

The transportational aspects of such near omnipotence, as we saw above, are not discarnate – our physical bodies accompany boats upon the water and planes in the sky. The same is true with spacecraft that take us off of this planet. But in as much as space vehicles carry not only people, but communications equipment such as video cameras, that triumph in transport became a triumph of communication, a triumph of and beyond our Earthly perspective.

A key moment occurred in the Apollo 17 moon trip – not the first, but sadly the last expedition to bring people to the moon, at least in the twentieth century. The video transmission from Apollo 17 on the moon was full-color and clear, much more sophisticated and vivid than the blurry black-and-white transmission of Neil Armstrong's extraordinary first step on the moon during Apollo 11. At one point in their Apollo 17 lunar sojourn, astronauts Gene Cernan and Jack Schmidt pointed their powerful video camera back at Earth. The significance of that instant was not lost on the control team back in Houston: so sharp and penetrating was the living image of Earth conveyed, with shore lines visible beneath the clouds, that the people in Houston could easily believe the camera from the moon was looking back at them (see *From the Earth to the Moon*, Part 12, HBO-TV,

1998, for more). They were processing an image from the moon that contained within it their very processing of that image, and everything else then taking place on Earth. You can't get much more godlike than that.

McLuhan traced the import of this extra-Earthly view to its very inception, observing (with Bruce Powers, 1989, pp. 97–8) an "an attitude of mind which has persuaded Western man to take on the duties of a god. Sputnik in encircling the planet made it an object of art. That small aluminum ball called forth a view of the Earth as something to be programmed." Whether this contradictory image was due to Powers or inherent in McLuhan – art and programming do not sit especially well together, and no one considers painting by numbers to be art – it captures the increase in control and responsibility that a vantage point beyond Earth bestows upon us (see Chapter 12 in the current book for more on Sputnik and Earth-as-art, and Chapter 13 for more on art in the digital age).

McLuhan and Powers in the same pages call this new omnipotent attitude "deadly...especially when those affected by it are unaware of its cause," but McLuhan also took a more positive lesson from the extraterrestrial vantage point, noting (with Barrington Nevitt, 1972, p. 7) that "the new ecological age...began with Sputnik." This notion of caring for our planet, and the connection of this notion to looking back at our planet from afar, was typified and most implemented by Stewart Brand's "Whole Earth movement," which began in the 1960s with his call for "a picture of the Whole Earth." Significantly, Brand's ecological concern by the 1980s had grown to encompass personal computers and the Internet as vehicles for human improvement – "Whole Earth" catalogs increasingly became catalogs of software (e.g., Brand, 1985) – a perspective with which I very much agree. Whether art form or program, the Earth and we can best be cared for with a maximum not a minimum of information technology.

Yet the deepest lessons of images beyond Earth are that technologies on their own are not enough. Discarnate man ironically cannot ply the void and bodies of outer space – we need physical transport for that, along with full flesh-and-blood people. Machines launched without people are often subject to error that can only be corrected by people. Thus, the Hubble telescope, which in allowing us to peer into the reaches of the universe to discover possible planets around distant stars has made us godlike in another way, was barely operational after its initial launch. Fortunately, a human being, Story Musgrave, was able to hop in a shuttle and fit the Hubble with corrective lenses. The general point is that technologies are intrinsically fallible – because we, their human creators, are imperfect –

and thus intrinsically unlikely to be self-correcting, especially when working at the cutting edges of our knowledge and capabilities. But we as living beings *are* inherently, if incompletely, self-correcting – that is what evolution is all about.

Indeed, evolution teaches us something else about discarnate humanity: we/they have been around a very long time, perhaps at the very origins of life itself. For what is DNA, but a discarnate code that looks nothing like the organisms and beings it commands into existence? Thus, we find the indispensable discarnate/incarnate relationship at both the beginning and future of the path we have travelled to the stars.

THE DEVIL IN THE DNA

Long before computers made us angels – in the Sunday school sense and otherwise – long before our television and movie stars and political leaders and sports heroes were everywhere at once in their images, DNA made the Earth its heaven by being everywhere at once on this planet. To be sure, there were and are differences in the DNA of different organisms, just as there are differences in angels and movie stars and political leaders. But an essential sameness in all DNA – which allows a bacterium genetically altered to produce human insulin, or the gene of a fish to be inserted into the strawberry genome to make its fruit more resistant to cold, for example (see Specter, 1998, for details) – prevails.

DNA is discarnate insofar as it is not itself a physical body – its relation to physical bodies is on this level no different than the relation of a computer program to the tasks it makes the computer perform. Yet just as software cannot do anything outside of an appropriate hardware environment, so too is DNA powerless, meaningless, without raw living material to shape into organisms. And here discarnate DNA goes beyond the software/hardware relationship, because living bodies are not only necessary for DNA to perform, they are the result of that performance.

Moving up the evolutionary scale, human thoughts in their relationship to brains partake of something similar to DNA and bodies. (I use the word "up" because I see the emergence of human thought as genuine, self-evident progress in evolution.) Thoughts are clearly discarnate, something other than material. The speed of imagination is the one known "force" faster than the speed of light – we can get from here to Alpha Centauri and back instantly, in contrast to more than eight years at the speed of light – but imagination can attain this subjective speed only because it need not perform in any physical universe. And yet, it needs a crucial piece of the physical universe – brains – in order to occur at all. Thus, all thoughts are what I call "transmaterial" (see Levinson, 1988b), in that they need some material substrate in which to exist – brains at first, later the air (for speech), paper (for writing), computers, etc. – and yet regularly and naturally go beyond those substrates in their travels (to Alpha Centauri or wherever – time travel is also possible in imagination), which they can accomplish as long as the travel stays virtual. Imagination, then, is much like discarnate life and sex in cyberspace: they are reliably satisfying only to the extent that they refrain from the corporeal.

But thought and DNA have something else in common: just as DNA can and does arrange proteins into living organisms, so too can thought reshape material. Indeed, the result of thought operating upon material is technology, which can be seen as the tangible embodiment of ideas, plans, schemes, and dreams. As I explore at length in *Mind at Large: Knowing in the Technological Age* (1988b), any given technology, such as the automobile, is both an embodiment of one central idea (in this case, to travel without being drawn by horses) as well as a compilation of numerous embodiments of other, different ideas (wheels to move, glass to let in light but not rain, etc.), each of which serves as a constituent of the overall technology, and is subsumed in its whole. The analogy to living organisms is, I think, striking, and coincides with earlier situations of technology in the biological world by thinkers ranging from Samuel Butler to Norbert Wiener (see Levinson, 1979a, for a summary). Donald T. Campbell's "evolutionary epistemology" (1974a) and the analogy it draws between the lives of ideas and organisms also would support a discarnate contribution to technological as well as living material.

McLuhan enjoyed and employed the biological analogy for technology too – most famously, as we saw in Chapter 3, in his observation (1964, p. 56) that "man becomes, as it were, the sex organs of the machine world, as the bee of the plant world, enabling it to fecundate and to evolve to ever new forms." The Butler crack from which this derived – that the chicken is the egg's way of making more eggs (Butler, 1878/1910, p. 134) – may be older than Butler, since he prefaces his observation with a significant, "It has, I believe, often been remarked that..." Dawkins' (1976) further update that the body is the gene's mechanism for propagating more genes, together with his parallel notion of "memes" – the idea as a virus or gene, which hijacks the human mind and persuades it to do its bidding, i.e., propagate the idea through speech and writing (as I am doing right here for the idea of meme, in this very passage of this book) – chimes nicely with both McLuhan's notion of humans as the pollinators of technology, and mine of technology as the embodiment (and thus propagator) of human ideas.

We thus come full circle to the sexual energy – the virtual incubi and succubi – of the online experience, as well as to the inevitable involvement of the physical body in any discarnate experience other than fantasy.

But the road from DNA to the stars is peopled with many things. When we're not making love – real or virtual via letters and machines – we're engaged in business, entertainment, politics, and all kinds of group affairs. To the extent that such activities, unlike sex, are not interpersonal, they flourish in mass media like radio and television, where we find the voice and face of the President and the movie star, the home-run hitter and rock 'n' roller, every place at once. They entail their own mix of ether and flesh. Although their senders are sent, most of us are only their receivers, and we partake of their virtuality only as listeners and viewers, with our creative mentalities as well as our bodies firmly kept on the tangible side of the divide.

This is a world, however, that the Internet is beginning to revolutionize.

It is a world that McLuhan noted so well in one of his best known – and best understood – taglines. It is the subject of our next chapter: the global village.

6

FROM VOYEUR TO PARTICIPANT

In a volume such as this, in which so many of McLuhan's ideas are inevitably described as "among his best known, least understood," it is refreshing to come upon the global village.

Introduced, like many of his key concepts, in his typescript "Report on Project in Understanding New Media" (1960, p. 129), and then to the world at large as a chapter title in *The Gutenberg Galaxy* (1962), McLuhan's suggestion that "The new electronic interdependence recreates the world in the image of a global village" (1962, p. 43) found a ready and comprehending audience. So popular was the phrase that it made its way into the title of two of McLuhan's subsequent books – *War and Peace in the Global Village* (McLuhan & Fiore, 1968), and his posthumous collaboration with Bruce Powers, *The Global Village* (1989) – and it has been bandied about regularly and more or less aptly in newspaper and magazine articles, in radio and TV commentary, for better than 30 years. No other of McLuhan's aphorisms has had it so good. Not "the medium is the message," which, although at least as well known as the global village, and punningly employed, again with Quentin Fiore, as the title of their *The Medium is the Massage* (1967), has been almost hopelessly misunderstood (see Chapter 3 of this book); nor McLuhan's "hot and cool," which had quite a ride in the 1960s and 70s, but is today almost forgotten as well as misunderstood (see Chapter 9 of this book for an attempt to reverse both of those last results).

The appeal of the global village as metaphor is easy to see. Once upon a time, denizens of the local village had more or less equal access to all public information – the voice of the town crier reached everyone. Print greatly expanded the reach of information – creating the first mass audiences, the first fast, big publics beyond eyesight and earshot – but also shattered the simultaneity of the original, acoustic, village crowd. Not everyone subscribes to the same morning or evening paper, and those who do are not likely to all read the same story at the same time. Enter radio, and then television: everyone sitting in their living rooms around the country hears the voice, sees the face of the speaker of news, at the same time. The village has been reconstituted, if not literally on a global level – that would await the advent of global cable news stations like CNN in the 1980s, and even then only partially – at least on the national level, or close enough to make the global village ring true as an effect of broadcast media.

And yet, at the time of its inception, the global village as metaphor could not ring totally true. It is in the nature of metaphors, remember, that they cannot be identical to the realities they seek to elucidate. Thus the village, in its original town-crier sense, was an information environment that allowed receivers of information to at any time become senders. Just as in a classroom, those in the village could ask the communicator – the teacher or crier – a question. Again, the loss of this possibility for immediate dialogue in written communication is what Socrates decried in the *Phaedrus*. And broadcast media were mute about rectifying this loss – the villages of listeners and viewers that these one-way mass media convened (and still convene) were comprised entirely of eavesdroppers and voyeurs, unable to ask Edward R. Murrow a question, whether on radio or TV.

This shortfall in the initial global village is important, precisely because it has been rectified to a growing extent on the Internet. This chapter might be considered the story of how the Internet made an honest metaphor out of the global village – or rather, converted it from a metaphor to something much closer to a depiction of reality.

First, we explore some of the characteristics of McLuhan's classic global village – the village of children and voyeurs.

VILLAGE OF CHILDREN, VILLAGE OF VOYEURS

The traditional family is unavoidably authoritarian in the relationship of parents towards children. Like the anecdote about Abraham Lincoln and his Cabinet, in which the entire Cabinet votes one way, Lincoln votes the opposite, and aptly says that his lone vote carries the day, so no number and intensity of opinions from children can be permitted to hold sway against an opposite opinion by parents on a crucial issue. Children may be encouraged to voice an opinion, just as teachers may elicit views from their students, but ultimately parents and teachers set the agenda. Thus, carried to the extreme, and depending upon how much democracy the parent wants to indulge (and this of course depends upon the age of the children, the issue at stake, etc.), one could say that in the parent/child relationship the voice of the child counts not at all. Hence, "children should be seen, not heard," from the parents' perspective – or, from the perspective of the children, should keep their ears open and their mouths closed.

The remarkable resemblance of this family dynamic to the environment of radio can be appreciated not only in its technology, which allowed people to hear without questioning, but its political impact. As discussed in more detail in *The Soft Edge* (Levinson, 1997b), the age of radio gave rise to the four most powerful political leaders of the twentieth century – Stalin, Hitler, Churchill, and Franklin Roosevelt – each of whom used the then-new medium to address their people at crucial moments in history. At these times, their voices came into living rooms, heretofore the precinct primarily of family, and spoke via a device unable to register any contrary opinion or objection. Listeners, whatever their ages, became children at the feet of these radio fathers. Their nations became not villages, because villagers can be contentious and democratic, but families, in which every citizen within earshot of the broadcast was a member without authority, a child. The level of political control this engendered and sustained of course was modulated by other factors – Churchill's party was voted out of office after the Second World War – but the national families constituted by radio even in the democracies were unique, with nothing quite like them either before or after.

McLuhan caught the fleeting reign of the radio family in his oft-quoted observation (1964, p. 261) that "had TV come first there would have been no Hitler." Indeed, the family-style "village" engendered by radio was soon to take another turn. By 1960, John F. Kennedy was admired at least as much for what he looked like as what he said. And the admiration was more of fans for a movie-star than of children for a father. His administration became Camelot, idolized, loved, enjoyed by the American people as they would a parade or a television series. Its tragic ending was heart-rending, but in the sense not of children losing a father – my parents often told me, about Roosevelt's death in 1945, that they felt as if they had lost a father – but of something more akin in distance and intensity to the death of a teenage crush. The global village was growing up, via television, from listeners to viewers, from children to voyeurs.

The object of the voyeur of course can be loathed as well as loved – hence our fascination with serial killers and horror movies – and the political leader in the age of television has been subject to both projections. American Presidents Lyndon Johnson, Richard Nixon, Gerald Ford, Jimmy Carter, and George Bush were either voted out of office after one term (or part of one term, in Ford's case), or obliged to leave for other reasons. Ronald Reagan, literally a movie star before he entered politics, was a two-term President and far more loved than the others, but even his administration was shaken by a crisis – televised, of course, via the Oliver North, Iran-Contra hearings – as if the American public with its thirst for political circus wanted to see at least some mauling of its heroes in the ring. And then came the only other two-term President thus far in the village of televised viewers, Bill Clinton…

If John F. Kennedy inaugurated the television presidency and its voyeurs – his close victory over Nixon widely attributed to his success in the Kennedy–Nixon debates, judged by a majority of radio listeners to have been "won" by Nixon, a majority of television viewers by Kennedy, with more people seeing the debates on TV than hearing them on radio in 1960 (see McLuhan, 1964, p. 261, and Levinson, 1997b, p. 90) – he managed to escape their most prurient interest. The escape was narrow. JFK, if posthumous reports of his numerous affairs with Judith Exner and others are to be believed, certainly offered more than enough raw material for a media feeding-frenzy. In England, the Profumo scandal was demonstrating just how evocative of public attention the mixture of sex and politics could be – far more than pro forma, for the media fumes from those events brought down part of a Cabinet. But the village of video voyeurs waited until the Presidential primaries of 1988 to wield its first nationally destructive impact in the United States, when the candidacy of Gary Hart was short-circuited by reports of an extra-marital affair. Moved by this event – I preferred Hart both to his Democratic rival, Michael Dukakis, and the Republican George Bush, who won the presidency that year – I published an essay online entitled "Can Only Angels be President?" (reprinted in Levinson, 1992).

The campaign of Bill Clinton for President in 1992 quickly ran into problems of the "Hart," and seemed destined to end in the same politically disastrous manner. Significantly, it did not, and Clinton was elected to office notwithstanding reports of an affair with Gennifer Flowers, which were denied by Clinton even as he alluded to having other unspecified extra-marital dalliances. Like so much else in the study of media and politics, this result admits to at least two rival interpretations: (a) Clinton was elected despite extra-marital activities because the village of voyeurs was growing tired of this particular spectacle, (b) Clinton was elected *because* of the extra-marital activities, because the village of voyeurs (or at least a majority) had come to enjoy them, rather than be upset by them, and the village wanted to insure that this form of entertainment continued. The second interpretation is a bit cynical, I admit, but in all likelihood both interpretations played a role in Clinton's victory in 1992.

Both would be put to the test again in 1998, when the Monica Lewinsky scandal erupted on the night of the Pope's visit to Cuba, and pre-empted that auspicious event in prime television coverage.

But by 1998, the television screen and the village of voyeurs it played to was no longer the only screen in town. It had been joined by a screen on people's desks in their homes and offices, a screen on their laps as they travelled in planes and trains – a screen which, when connected by wires or remotely to a telephone system, allowed people for the first time in the history of screens to do more than just view them.

These screens are of course the windows of the Internet. Their supplantation of both television and newspapers as the cutting edge of news was signalled by the Congressional release of the subsequent Starr Report on Clinton's extra-marital sexual activities (Office of the Independent Counsel, 1998) directly to the World Wide Web, to which the public and the media enjoy equal access. (An earlier example of international news reported by the government directly on the Internet was the release online by a Massachusetts court of its decision to reduce the sentence of Louise Woodward in the "au pair" case in 1997.)

By the end of the twentieth century, McLuhan's global village was turning, year by year, into a more full-fledged reality.

POLITICS IN THE ONLINE GLOBAL VILLAGE

Ever since I was a kid, and ever since I began reading science fiction and watching it on television – all three periods coincide for me, in the early 1950s – I have been hearing about democracy through computers. The idea was that, rather than electing representatives – representatives and senators in the US, members of Parliament in the UK, etc. – to thrash out issues via discussion and vote upon them on our behalf, we the people could discuss the issues and vote upon them ourselves, through "the computer." I was not sure, at first, just how that computer would work, and how it could be reached. But I came to understand that it could tally votes, and could be reached in some way via the telephone – with operators who could be called, who would in turn enter the votes into some big central computer – or maybe by terminals in libraries, schools, town halls, the same public places in which we cast our votes to elect representatives.

In the past two decades – or since the advent of personal computers and modems – I came to realize how much easier such a direct voting procedure could in fact be: we could cast our votes directly from our homes and/or places of business, via the computers and modems which have become almost ubiquitous in our society.

This, then, is the realization, technologically – or the potential for realization, if we choose to implement such a process – of the political dimension of McLuhan's global village. There is little if anything metaphoric about it. From the perspective of the information that in-person ballots cast, there is nothing different or lost in a ballot cast online. Like electronic fund transfer versus handing a ten-dollar bill to someone, there is no advantage to the transaction in conducting it in the flesh (although we might enjoy and thus prefer an in-person exchange or in-person ballot for other reasons, extraneous to the transaction, such as we find the bank teller or the person standing on line in front of us very attractive; but, then again, we might not).

Further, the history of democracy speaks for the legitimacy of direct voting via computers. The ideal size of the Athenian democratic state – whose legislators were not elected representatives but every citizen – was defined as the number of people who could gather to hear a speaker. In such circumstances, citizens could both ask the speaker questions and confer among themselves. Although the mass media of the press, radio, and TV expanded that audience enormously, to national levels, it did nothing at those levels to allow the audience to speak back. Nor did it even allow the members of the audience to speak to one another, except in very small units like families and gatherings at the pub and conversations with a friend, which amounted to minute slivers of the national audience. Thus, the mass media in effect made direct democracy impossible, even as the press, especially, became the indispensable Jeffersonian bulwark of triumphant representative democracies in America and Europe in the last two centuries. Lewis Mumford aptly captured the price of this triumph when he testified to the U.S. Congress in the 1960s that "democracy, in any active sense, begins and ends in communities small enough for their members to meet face to face" (quoted in Will, 1998, p. 6). But, of course, in an age of one-way, non-interactive mass media, we had no other choice.

McLuhan was quick to see, even in the age of one-way television, how new media could make large areas small enough for that kind of interaction. "The tribal will," he said in his *Playboy* interview (1969, p. 72), "is consensually expressed through the simultaneous interplay of all members of a community that is deeply interrelated and involved, and would thus consider the casting of a 'private' ballot in a shrouded polling booth a ludicrous anachronism." The ballot McLuhan is referring to, of course, is for election of representatives to debate and pass laws, and make decisions on our behalf.

Three decades after McLuhan's observations, online communication has begun to open some avenues to global implementation of Athenian local democracy – or at least, national expression of the directly democratic city-state. The Internet already provides numerous places and occasions for discussions, some including questioning of leaders in various sectors, if they choose to make themselves available online; software for achieving and gauging consensus, as well as casting and tallying of votes, public or private, is readily available (voting options have been featured on specialized online systems since the late 1970s; see Hiltz & Turoff, 1978/1993, and Stevens, 1987, for details).

The question is: given that we may have the technological wherewithal, do we want to use it?

Walter Lippmann's *The Phantom Public* (1927) – published, appropriately enough, at the dawn of broadcast radio and its maximized environment of speechless listeners – offers the blunt argument against direct democracy. "We must abandon the notion that democratic government can be the direct expression of the will of the people," Lippmann urges. "Instead we must adopt the theory that...people support or oppose the individuals who actually govern." Such indirect, representative government is necessary, Lippmann explains, because the affairs of state and society "are altogether too numerous, too complicated, too obscure in their effects" to be comprehended by the private citizen.

Of course, Lippmann's public was indeed phantom, bereft of its capacity to question Lippmann, its leaders, or even to converse internally about national issues except in the smallest local aggregates. Nor did radio and television, as we have seen, help the matter – rendering only leaders and stars, not members of the general public, discernible on millions of boxes in the ensuing decades.

But with millions now actively conversing on the Internet, with the private citizen no longer Lippmann's "deaf spectator in the back row," but instead able to access information about most events more easily than most senators in 1927, have the affairs of the world become more comprehensible and hence governable by the average person?

Whether Lippmann would have thought so is not clear. Although he stops short of a Platonic dismissal of the common person as in effect just too stupid to govern, he makes a point of emphasizing that the public cannot be educated to self-govern. His reasoning comes straight from the mainstream of early-twentieth-century American philosophy à la John Dewey, and its insistence that first-hand experience is the best and even the only valid teacher. People on this view cannot be taught to govern, certainly not in formal schools and likely not by the dissemination of any amount of information about governing; they can only learn how to govern by being on "the inside," according to Lippmann, i.e., by being a member of Parliament or Congress already.

But this need not slam the door on self-governance via the Internet. For whether or not a present-day Lippmann might be correct that dissemination of information in the online global village would be insufficient, in principle, to make us legislators, the global village is more than just a means of dissemination: to the degree that it is also a means of discussion, debate, consensus-building, and voting, it is therein in itself a mechanism of government. When writing about the Communications Decency Act (1996) for *The Soft Edge* (1997b), I was able within minutes to locate not only the full text of the law, but also earlier drafts, and lengthy commentaries and critiques of the law by numerous citizens. And in the months that followed, I participated in many an informed discussion about the CDA in many places online. In what way was I, and other parties to these online discussions, less "experienced" in the workings of government pertinent to this law than the Congress that considered and approved the law? (I would contend, in fact, that I and the many other people online – who immediately denounced the law for the unconstitutional incursion on freedom of speech and press that it was – were apparently far better at governance in this case than either Congress or President Clinton who signed the bill into law; in July 1997, the Supreme Court struck down the law as unconstitutional.)

Of course, this was but a single law. One might well argue that people who discussed the CDA online were a self-selected group who already had interest in and knowledge of this area – that Congress in any year considers a wide array of laws, far beyond the capacity of citizens even enabled by the Internet to meaningfully assess and discuss. This argument focuses especially on Lippmann's contention that the issues of government are "too numerous" for the public to constructively engage, and resonates with the general hue-and-cry in our own digital age about "information overload."

I have found that this problem succumbs to a one-two punch that begins in the philosophy of William James, a colleague of John Dewey in pragmatism and the primacy of experience. For James, the world is a "big, booming, buzzing confusion," until we apply our mind to it, and in so doing sort things out – select from the background, into the foreground, those aspects of the environment most pertinent to our proximate needs and experience. Human mentality, in other words, is for James in the very business of processing information overload, and wringing from it meanings appropriate to the tasks that confront us.

Since our cognition thus comes already equipped to deal with overload, worrying about it is much like having anxiety about not having enough air to breathe when our lungs are working fine. The anxiety itself – not the perceived lack of air or the purported surfeit of information – is the real culprit.

The second punch that beats this problem arrives in the evidence of how well we deal with presumed problems of overload outside of the digital, virtual arena. Even the smallest bookstore or library contains many more volumes, choices of texts, than we can possibly consider. But rather than breaking down in a paralysis of apprehension about such overload of choice, we instead cope quite well and sooner or later locate if not the exact book we were seeking, at least a reasonable alternative. We are able to obtain such satisfactory results by applying a navigational system – about where books are shelved in bookstores and libraries, such as biographies under the names of their subjects rather than authors – that we have internalized in childhood. Such systems are apparently quite capable of being taught, formally and informally, and together with James' organizing mind they fare very well for us in our attempts to have some control over our lives and this world.

Today the digital world indeed presents a far greater number of choices than even the biggest bookstore or library. But that world also is increasingly providing us with new navigational systems – ever more sophisticated search engines – to help us wring data pertinent to our tasks at hand from the vast online array. Indeed, to the extent that search engines are becoming more "intelligent" – i.e., they can be programmed to act as our agents, to do our specific bidding, to keep their "eyes and ears" open for information that we have earlier flagged for them – such engines are even more consonant to the workings of the human mind than the "Dewey decimal" card catalog system (developed by a different Dewey – Melvil – who did live at the same time as John) or the shelving policies of bookstores.

To be sure, none of this is any guarantee that government by direct democracy, empowered by access to information, discussion, and voting online, will work, or work any better, than our current representative kind. The people of ancient Athens did have somewhat fewer issues to consider than we do in our modern world; and for all we know, the complexity and sheer number of such issues does require a certain talent in governing. It may also be that direct democracy might work well for only some aspects of government – perhaps better for the local than the national (and international) legislative after all.

On the other hand, if the performance of our representative democracies, and the satisfaction of the people in their performance, are any indication in the past few hundred years, we might well try to implement some of the directly democratic opportunities of the interactive, online global village. Although what Churchill called the "least worst" form of government has been spectacularly superior to its authoritarian and totalitarian alternatives, why not try for something even better – a system worthy of three not just *Two Cheers for Democracy* by E. M. Forster?

In the meantime, the online world has already begun to function as a full-fledged village for activities other than politics, including business. Of course, as with politics, we will see that, in the global village, it is not quite business as usual…

BUSINESS IN THE ONLINE GLOBAL VILLAGE

On the one hand, radio and television – the bulwarks of McLuhan's classic, i.e., non-interactive global village – were businesses literally writ (and shouted) large, at least in the United States, where advertising upon them became and still is an art form, a cultural phenomenon, and a huge industry. On the other hand, with the small exception of direct sales made via the telecast (this exception has grown a bit with QVC, the Shopping Channel, and similar cable TV stations in the US), actual money exchanged hands neither between audiences at home and the people on radio and television, nor among members of the audiences themselves. Thus, in that sense, the "villages" of radio and TV – the villages, that is, comprised of their listening and viewing audiences – did no business at all.

Advertising as the essential business of mass media began with the printing press, and gave it the economic nourishment to function independent of government and its funding. In this symbiotic relationship crucial to democracy, advertisers gained a larger and more informed public for its products, and newspapers were able in the rest of their pages to print what they wanted. The Jeffersonian view of the press as the fourth branch of government – exerting an important check on the performance of the other three in the United States by being ever-ready to criticize them – would have been unworkable without a means of support for the press removed from government. Democracies in England, France, and other parts of the world have been likewise well served by a free press made possible by advertising.

For a variety of reasons, however, only American broadcasters wholly adopted the advertising/newspaper symbiotic relationship – England, France, and other democratic societies for most of the twentieth century instead saw broadcasting as a form of postal service, or a vehicle of communication so vital that it could not be entrusted in any way to advertisers. (The American Jeffersonian view, in contrast, is that broadcasting is so vital a form of communication that it *cannot* be entrusted to government; nonetheless, the passage of the Federal Communications Act in 1934 in the United States let in the government via a back door anyway; see Levinson, 1997b, pp. 84, 154, for some of the continuing deleterious results.)

In the past few decades, many countries have begun to allow various forms of advertising into their broadcasts. Ironically, a socialist regime initiated private ownership and funding of radio and television in France (see Head & Sterling, 1987, p. 21).

Advertising has played a growing role on the Internet as well – indeed, its presence there has been a subject of some controversy – and yet the business transacted online has been from its outset far more direct than the three-way commerce of advertiser pays communications medium, which displays the ad to the potential consumer, who the advertiser hopes will become a real consumer of the advertiser's products. Instead, Internet users who access Amazon.com, for example, can buy books as directly from Amazon.com as they might when entering a Borders bookstore in person. And the business of the Internet is direct in another way: although most of the online sites that one surfs to and through are free, the initial Internet access – the capacity to log on to the Web via a browser in the first place – usually is not, unless the surfer is a student in a university, or a visitor in a public library. Thus, Internet users pay subscription fees to Internet providers, much as consumers subscribe to magazines, telephone services, and cable TV systems – unlike the audiences of broadcast radio and television, who are voyeurs economically as well as informationally, in that they have no direct economic interaction or relationship with the broadcasters.

The most radical online business of all, however, and a commerce which most completely fulfills the ideal of a virtual online community, would be a business in which the products as well as the payment were provided entirely online – in which not only the service rendered, but the money tendered, was virtual or wholly contained in the cyberspace village. By this standard, all current business in the online village is at most a partial implementation. Amazon.com sells the same books available in physical bookstores – not hypertexts or other online texts – and is thus an online rendition of a catalog more than a village, whose communal aspects are expressed more in the reviews posted by online readers than in the books sold. Online education, in which students pay tuition for courses conducted entirely via personal computer and modem (see Levinson, 1995b, for details), comes closer to the ideal. But when online courses are offered for academic credit, it is granted by traditional, "place-based" institutions. Pornography offered on the Web may come even closer, and is thought by many observers to thus far be the only reliable way of making money online (Ferrell, 1996). But even in those hot cases, access is paid for by credit cards supported, of course, by traditional, stodgy banks.

Does this mean that the ideal of a self-contained global village – a virtual biosphere of business – is unattainable, much as the physical

Biosphere II in Arizona proved not feasible as a living community totally separated from the world? Must some crucial element of the business equation – a product, an academic credit, a credit card – always be grounded in the offline world?

Not necessarily.

At the outset of the online age, Murray Turoff proposed a virtual banking system that speaks to this issue (see Turoff, 1985, and Hiltz & Turoff, 1978/1993, pp. 466–7, for slightly later renditions of this proposal). He suggested that all members of an online village put x amount of real funds into an online bank, which would issue to each member an equivalent amount of online currency. Those virtual funds – redeemable at any time in real money – could then be used for any online activities.

Here is how it might work:

Let's say I run an online seminar about Marshall McLuhan, and those who enroll provide payment to me in virtual currency. One of my students is a talented organic gardener; I seek her advice on how to improve the tomato yield in my back yard, and pay for that advice with money underwritten by the First Bank of Cyberspace – i.e., the people in charge of keeping track of the virtual/real currency exchange.

Of course, even in that scenario, money is ultimately rooted in the real world, just as my tomatoes, and Amazon.com's books. But it is a step towards a more self-sufficient interactive virtual village, and it could become even more virtual via the replacement of money altogether with barter. Thus, rather than charging even virtual tuition for my McLuhan Seminar, I accept from my students a pre-set number of barter vouchers. Then, when I seek advice from the garden expert about my tomatoes, I in turn pay her with vouchers. Obviously, for such an economic community to operate, its members must have information and/or services to sell – of mutual interest and perceived value – and the denizens need to stay online for a period of time sufficient to use and benefit from the barter vouchers they have amassed. A good consequence of that, if we are attempting to progress towards more self-sufficient online villages – or perhaps even a truly global village of business – is that people are encouraged to continue in the online community.

At present, however, virtual money in any form other than what can be immediately converted to hard cash or its equivalent is still virtual in the ancient sense of the word – i.e., as far as I know, it exists only in the realm of suggestion and theory (or, if not, in some limited experimental applications, which as yet have not enjoyed widespread use). Indeed, even

Theodor Nelson's (1980/1990) more modest proposal for, in effect, an online cash register attached to every document on the Internet, such that each time a reader accessed the document the cash register would ring up the sale by electronic fund debits from the reader's offline account, remains unimplemented.

The impediment, as with direct democracy in the global village, is not technological. It is rather a problem of attitude (just as the possibility of direct democracy clashes with the attitude that people cannot govern themselves without representatives) – in this economic instance, an attitude that wants everything on the Internet to be free.

As I explored at some length in *The Soft Edge* (1997b, chapter 17), this attitude begins with an equation of all virtual activity as a form of informational property (fair enough), which in turn is seen as being significantly different from tangible property or real estate and services, literally (still fair enough), with the conclusion that informational property (also known as intellectual property) ought not to be bought and sold. That conclusion, I argue, is not only unfair but unwise, in that it punishes rather than rewards the creators of information.

Interestingly, the "information wants to be free" anthem is most traditionally associated with Stewart Brand (see Barlow, 1994, for details) – more recently Esther Dyson (1997) and others – the same Brand whose "Whole Earth" vision was an early expression of discarnate humanity, a view of Earth from a position beyond it (see the previous chapter), and who regularly and appropriately acknowledges McLuhan as an intellectual guide (see, for example, Brand's quote from McLuhan as the frontispiece of the first chapter in Brand's *The Media Lab*, 1987, p. 3).

And what does McLuhan say about this?

He is apparently not a forebear of Brand in the latter's view that information wants to be free.

Indeed, in one of his most insightful chapters in *Understanding Media* (1964) – "Money, the Poor Man's Credit Card" – McLuhan characteristically charts the quintessentially human impulse to trade, purchase, obtain, compensate, from its pre-literate vessels (barter) through its alphabetic/literate expressions (coinage to money printed on paper) to its post-alphabetic, electronic forms. Significantly, although these forms are not money – they are credit and credit cards, apparent as money's successors even in 1964 – they are by no means an abnegation of property and purchase, as urged by Brand, Dyson, et al. at least regarding information and virtual activities. Indeed, closer to our own time, and well after the first rush of online communication in the 1980s, McLuhan (posthumously) and Bruce Powers repeatedly emphasize the replacement of money with credit in the digital age, the similarities of credit to barter, and the general continuance of economies, albeit in very different forms (McLuhan & Powers, 1989).

Thus, although McLuhan correctly identified copyright as a product of the printing press and its fostering of "the habit of considering intellectual labor as private property" (McLuhan & Fiore, 1967, n.p.), he never said that intellectual work should be provided without compensation, nor did he see the world after print as incompatible with the buying and selling of services and products of the human mind.

To be clear, he saw the electronic age, and the advent of electronic credit in particular, as eliminating many things – such as the privacy of finances, impossible in an age in which credit references and records of anyone can be had with a quick dance of someone else's fingers across an appropriately connected keyboard. But finances themselves – intellectual and informational property and its obligations – were not among the eradicated.

The global village as a functioning self-sufficient commercial entity, then, at least on McLuhan's reading of the electronic age and its impact on economics, remains an intriguing possibility, if not yet anything close to an actuality. And in the meantime, a hybrid global village of commerce digital and analog – comprised of online transactions for online and offline items and services, all financed by offline lucre and credit – becomes a more literal description of our world every day.

This is not surprising, since, as McLuhan often noted and we will consider in more detail later in this book (see Chapter 14) "we march backwards into the future" (McLuhan & Fiore, 1967, n.p.), clutching on to as much as possible of the known past as security. Academic credit from brick-and-mortar institutions, books of paper, banks we can walk into to obtain our money are all, whatever else their advantages, necessary stepping stones to futures already emerging in which people take courses for knowledge not credit, books can be read on screens, and money likewise can be earned and spent without ever being held in the hand as cash, check, or credit card.

But if we have seen in this chapter that the notion of village is not incompatible with such virtual renditions – indeed, they help bring communities into being – we will see in the next chapter that the great cities of the twentieth century, according to McLuhan, stand very much at risk in this digital age.

7

THE FATE OF THE CENTER

So crucial to the well-being of humanity did the Roman Catholic Church hold the notion of the center – not necessarily the Earth as center, but something, somewhere, as center – that they burned Giordano Bruno at the stake in 1600. The Copernican challenge to Ptolemy and his Alexandrian placement of the Earth at the center of the universe was disruptive enough. Cardinals Bellarmine, Barberini (later Pope Urban VIII) and others pressured, reasoned with, cajoled Galileo to recant his conclusions, based on new astronomical observations, that the sun not the Earth was the center – which Galileo did, his recantation a Pyrrhic victory for the Church in any case, given the continued outpouring of Galileo's books and their unrecanted arguments from the presses of Europe. But execution, as far as we can see in history, was never a real possibility for Galileo. The thrust of his "crime," though serious – substitution of one center for another is always unsettling – never endangered the true heart of belief. In contrast, Bruno contended that not only was the Earth no center of the universe, neither was our sun. He went far beyond the Copernican heliocentric theory supported by Galileo, and wondered if the myriad stars themselves might be suns, with planets around them, and thus each and every one its own center. A universe with an infinity of centers is of course a universe with no centers. Bruno's refusal to recant – aggravated in the Church's eyes by the fact that Bruno was a monk, rendering his heresy traitorous – was the proximate occasion for his execution. But the real target of his dispatchment was his utterly disconcerting, yet tantalizingly reasonable, idea.

Nowadays, as we embark into a new millennium, we not only find Bruno's idea reasonable, but data in support of it in the images of distant suns relayed to us via the Hubble telescope. Nor do we find it especially unsettling – perhaps because we have shifted our search for centers to investigations and theories of just where in the universe did the Big Bang, if that's what it was, commence. Or perhaps we are more comfortable now in a centerless universe because we increasingly live in a centerless world, from the point of view of information. There still are far fewer screens on this planet than stars in the universe, but each time I sit at a computer screen connected to the Web, I am seated at a center with far more planets of data than are likely to be revolving around any star.

McLuhan saw the beginnings of this centrifugal process, and began to gauge its effects, in the ages of print, radio, and television. "The city no longer exists," he wrote (McLuhan & Parker, 1969, p. 12) with typical extravagance but undeniable hint of truth, "except as a cultural ghost for tourists. Any highway eatery with its TV set, newspaper, and magazine is as cosmopolitan as New York or Paris."

Well, not quite. For New York and Paris, even today, when personal computers from anywhere in the world provide better access to some kinds of information than do the biggest urban libraries, still offer food for the intellect – and the palate, and other senses – unattainable in lesser physical centers.

But that's the end, or current terminus, of our story.

First, to see how we arrived here, we will trace the growth of centers everywhere, from the mass media that McLuhan scrutinized to the new media that have arisen in the decades since his death.

GOD'S SPACE

The notion of centers everywhere obviously relates to the global village, discarnate man, and, for that matter, acoustic space. McLuhan's "holographic" mode of thinking and writing pertained not only to his presentation of chapters within books – such as the 107 equidistant pieces in *The Gutenberg Galaxy* (1962) – but to his development of ideas across books, including those written with colleagues, and those written by erstwhile colleagues, such as Edmund Carpenter. Indeed, these ideas, and their dispersion like stars across the space of all of those pages, in themselves comprise a document of centers everywhere. *Digital McLuhan* can be considered a captioned snapshot of part of that document.

Carpenter (1972/1973, p. 3) makes explicit the relationship of discarnate man to omni-centrality. After noting that, "Nixon on TV is everywhere at once. That is the neo-Platonic definition of God" (see Chapter 5 in this book for more), Carpenter explains that God is "a Being whose center is everywhere, whose borders are nowhere." McLuhan makes an equivalent observation, more humanized and harkening back to an earlier broadcast medium, radio, when he relates how "Lowell Thomas used to say, 'On the air, you're everywhere'" (McLuhan & Powers, 1989, p. 70). The point is that, at least for St. Augustine, radio, and television, to be without a physical body is also to be everywhere at once.

The invocation of God is especially apt for the broadcasting of centers everywhere, as in radio and TV, in contrast to the more profound decentralization of personal computers and online networks. From the vantage point of the broadcast listener or viewer, who can hear and see anything with equal ease from any place in a given radius, the broadcast does indeed have a point of origin – or a center. And this center plays a role not only in terms of the technology – if one drives too far from the origin of a broadcast, its reception will be lost (unless a relay or network has been set up) – but even more significantly in determining the content of the broadcast. Thus, unlike the effectively unlimited options for content available to the Web browser, programming choices are sharply restricted for even today's viewer of cable TV in the United States. Moreover, in contrast to Web pages which are daily created and revised by myriad members of the Web audience, the content of radio and TV is produced by a relatively minute number of professionals, who work in stark superiority to their audiences. The notion of no centers is thus, for the broadcast milieu, as much an illusion as a reality.

We could examine the history of mass media in the last quarter of the twentieth century, however, and note how, even outside of the personal computer revolution, radio and especially television have become less centralized. Whereas in the United States there were once three major TV networks, by the end of the 1990s there were six. Whereas the overwhelming majority of TV viewers until the late 1980s watched one of the networks, viewership has since then been increasingly accorded to cable TV stations. And ever since the introduction of the video-cassette recorder in the mid-1970s, a growing number of viewers watch video tapes, of their own selection, on their TV screens, instead of a program supplied by a central network or station. (The advent of "call-in" radio and TV, in which listeners get to ask questions on the air, constitutes an additional empowerment of decentralized audiences.)

Yet none of these flickering screens and crackling speakers holds a candle to the Internet in terms of being everywhere at once – not just for the passive receipt of information, but for the active hunting, and the sending.

ONLINE EDUCATION EVERYWHERE

The online classroom serves as an ideal anteroom for the Internet, and its revolution in decentralization in comparison to the mass media of radio and television. On the one hand, as we discussed in the previous chapter, courses taken online for academic credit are still essentially tied to the central institutions – universities, state boards of regents, etc. – that grant the credit. In this sense, they still have a parlor contiguous with central authority. On the other hand, everything else about online courses is decentralized to a degree that far exceeds McLuhan's descriptions of mass media.

The 1980s marked the first continuous offering of online courses. The Western Behavioral Sciences Institute under Richard Farson began offering online courses as executive seminars (no academic credit) in 1982; the New York Institute of Technology with Ward Deutschman began doing the same with online courses for undergraduate credit in 1985; and Connected Education began offering online courses for graduate credit in cooperation with the New School for Social Research in 1985, with a complete online MA in Media Studies in 1988. I learned about the rudiments of online teaching at WBSI (for which I taught my first online course in 1984), and created Connected Education; my wife Tina Vozick and I still administer an online MA in Creative Writing via Connect Ed for Bath Spa University College in the United Kingdom. Thus, my experience in online education is first-hand.

The academic credit and MA in Media Studies degree in Connect Ed programs from 1985 to 1995 was granted by the New School for Social Research, itself a pioneering institution in providing courses for adult learners since 1919. But, with the exception of students already in New York City and in attendance at the New School who chose to take online courses, few Connect Ed students ever set foot in the New School, and those few who did came mostly out of curiosity – as if visiting the site of a printing press or a broadcast tower – not obligation. They came, in other words, and apropos McLuhan's view of the city in the electronic age, as tourists.

Their work as students was instead conducted from their homes and places of business in some forty states around the United States, and twenty countries around the world, including England, China, Japan, Singapore, Canada, and others in Africa, the Middle East, South America, and continental Europe (see Levinson, 1995b, for details). The teachers were also distributed, both across the United States, and in places ranging from South Africa to the former Soviet Union. Although the specific locales of students and teachers were of course of interest and added flavor to the courses, place in the physical sense of determining who could have access to the "classroom" was truly irrelevant in this online community.

Online education, with centers anywhere and everywhere a personal computer with a modem could be plugged into a telephone outlet, thus reversed a trend in higher education which had been developing continuously since the University of Paris, Oxford University, Cambridge University, and like institutions in the eleventh century. As the *Encyclopedia Britannica* (Grave, 1954, vol. 4, p. 652) aptly puts it about the earliest days of Cambridge, "it taught all comers who could live in the town and pay their lecture fees." Today, of course, these and other top-notch universities have rigorous entrance requirements; but pursuing a course of study still requires physically attending the university – living in its town. Indeed, in this insistence on place – one comes to Harvard to be a Harvard student (or professor) – our current institutions are one with a practice that not only stretches back to the origin of the university in early medieval times, but well before it to smaller schools like Plato's Academy and Aristotle's Lyceum. Although written works by both philosophers were available – albeit in numbers far fewer than after the printing press – students still had to study in the philosophers' physical proximity to receive the best education. And, given the fact that the written word, at that time, provided (as Socrates noticed) but one unvarying answer to all questions, a strategy of seeking education from a living, talking human being made good sense.

The dialogue made possible by online communication was thus an improvement, in some important respects, over both the place in education, and its faithful sidekick the book. Indeed, in as much as the online course entails lectures from faculty, questions from students and responses from faculty and students, discussion between faculty and students and among students, completion and evaluation of assignments – all of which are written and read as are books – the online course in effect is both a new kind of classroom and a new kind of book, a Hegelian synthesis of many of their best qualities which surpasses the classroom or the book operating alone, or even in tandem, offline.

But the advantages of centerless online education, though formidable, are complete over neither the classroom nor the book. On the one hand, the online class allows access twenty-four hours a day, from anywhere in the world. There is no limit on how many students can ask questions at one time – for unlike speech, questions in writing appear ipso facto in some kind of sequence – nor is there a limit on how many questions any given student can pose. Further, comments entered in an online course three weeks ago are as accessible as comments entered three hours ago; the asynchronous procedure of most courses allows students and faculty to log on at times of their own choosing, which increases the likelihood of a good engagement; and students can easily read each other's assignments after the faculty has graded them, and learn a lot from that. But notwithstanding all of these benefits and more, the place-based university still has that comforting presence of bodies; the psychological electricity (for the teacher) of walking into a classroom of people and having them hang on to your every word; the equivalent electricity (for the student) in so hanging on to such words, and for a moment or longer commanding that flow of information with a question; etc. The most incandescent online course pales in comparison to the pleasure of holding a class on the lawn on a sunny spring day, at least on the criterion of warmth on the skin, soft breeze in the face, to accompany the information. And the traditional offline book has kinaesthetic benefits too – the ease of holding it in your hand, with the rest of your body in a chair, a hammock, against a tree, on a blanket – which at present are lacking in most online experiences (see Levinson, 1997a, for more on the benefits and drawbacks of online education, and Levinson, 1998a, for more on the future of digital books; see also Chapter 10 in this book).

Of course, the ongoing evolution of media means that tallies of pros and cons are ever subject to change. Online books will no doubt be more comfortable to read as screens become less expensive; online courses already have the capacity to convey images of faculty and students if those are desired. But the above suggests that, notwithstanding the enormous advantages of participating in a university via a computer screen – notwithstanding the greater vibrancy, in many crucial respects, of online learning communities – the place will continue to have a place, at least in education.

But place is not the only enduring counterweight to the centerless society. Social attitudes, political institutions, themselves have weight, which can count against the acceleration of decentralization.

Indeed, in the case of central governments, they not only can count against but actively work against the emergence of centers everywhere and thus nowhere...

THE GRAVITY OF GOVERNMENT

The tendency of governments to reign by reining in decentralizing media was already apparent in the attempts of monarchs to control the first printers. As we saw in the last chapter, those attempts were ultimately defeated in democratic societies by advertising, which provided the press with non-governmental funding. In the twentieth century, broadcasters have been an especially appealing target of governments, since the sources of the messages that go everywhere are already so intrinsically centralized. Thus, we have become accustomed to hearing, in times of coups, that the revolutionaries have taken over a particular radio station, or the government is still doggedly holding on to it; when the last broadcasting centers have fallen to the rebels, we know that, whatever else may happen in the fighting on the ground, the revolution is won. It was won in the air – and the decisive victory concerned the origin of electro-magnetic carrier waves, not necessarily aircraft.

But what happens to the government's attempt to seize, or maintain, informational power when it is distributed in millions of computers, a significant portion of which act not only as receivers but as producers of Web pages, hosts of Web sites, in short, disparate centers of production and broadcast as well as vehicles of reading, listening, and viewing?

A calming point about the above is that governments in general have been rather unsuccessful in information control – thus the failure of monarchies to subsume the early printers – and totalitarian governments in particular have been rather lame as information dictators, despite their grim and horrendous efficiency in murder. Thus, Nazi Germany in the early days of the Second World War was unable to control the "White Rose," and the truth it distributed to some of the German population via primitive photocopying methods; in the end, the Nazis lost the war, to a significant extent because the Allies (Alan Turing, in particular) were able to crack the German "Enigma" code, thereby giving the Allies knowledge of all secret German operations. Germany's deficiencies in information technology were thus in part responsible for its defeat, despite its advantages in some important military technology, including rocketry.

And the collapse of the Soviet Union some forty-five years later was a conclusion of a similar story: despite its equivalence to and at times even surpassing of the West in military and missile technology, the Soviet Union was at a loss to compete with the United States in even the prototype phases of the Strategic Defense Initiative, and its reliance on sophisticated computer programming. Indeed, even at the height of its internal power, the Soviet Union was unable to control the thriving underground market of ideas dispensed via samizdat video and photocopying. Like the Thousand Year Reich, the Soviet Age fell victim to its informational incapacities. Contrary to George Orwell and the nightmare of *1984* (1948/1949), it seems that the more totalitarian the government, the less able it is to control information.

Nonetheless, even citizens of democracies cannot take too much comfort in such lessons, because their governments, as do all governments, have the power to heavily fine and even imprison purveyors and consumers of information not to the government's liking. Indeed, in the United States, which yet adheres to the death penalty for some crimes, the stakes are even higher. Joe Shea, prosecuted by the Clinton Administration under the Communications Decency Act of 1996, stood only to lose a few hundred thousand dollars and two years of his life to prison, if found guilty under the Act for publishing a letter criticizing Congress (for passing the Act) in colorful language on a Web page in principle accessible to children. But violators of the even more outrageous Sedition Act of 1798 could have been sentenced to death as traitors for some informational "crimes."

Fortunately, the Sedition Act expired with Jefferson's assumption of the Presidency in 1801, and the Communications Decency Act was struck down by the Supreme Court of the United States in 1997. But the Court's ruling was narrow – in effect viewing the CDA's attempt to regulate the Internet as an unconstitutional incursion on the Internet's performance as an online newspaper – and the door was left open for governmental attempts to control the Internet in less sweeping ways. Indeed, the Supreme Court throughout the twentieth century repeatedly endorsed the government's regulation of broadcast media, even though they too function as an arm of the press.

The American government's chronic misunderstanding of the Bill of Rights in the past century, especially as it relates to the flow of information, was brought home in a commencement address given at New York

University by Vice President Al Gore in May, 1998, in which he urged that government take steps to prevent invasions of privacy in the information age (see also Broder, 1998). He was referring to financial and personal data, compiled by credit bureaus and similar organizations, readily available about everyone to anyone with the dollars to spend on the fee. Certainly, control of such activities would be highly commendable. But the metaphor chosen by Mr. Gore to highlight the need for such controls was disturbing: what we need is an "electronic bill of rights," he said, to safeguard our privacy. Our Bill of Rights, however, is supposed to safeguard our rights of expression and dissemination – the free flow of information, as per the First Amendment, our rights *to* communicate, to send and receive – not protect us *from* expression and dissemination of ideas, however much such dissemination may compromise our privacy. Calling for an "electronic bill of rights" to protect our privacy is like calling for a Declaration of Independence to protect us from invasion.

And yet, the government's attempt to shift the object of our protections – the source of danger to our freedoms – from central government (as in the actual Bill of Rights) to central business (Gore's "electronic bill of rights," supposed to protect us from the rapaciousness of businesses compiling, storing, and selling our credit histories, etc.) is consistent with a society in which centrality is under attack by the Internet on many fronts. As the resource of centrality diminishes, those institutions that possess what little is left squabble to maintain their share, and increase it against the tide if at all possible.

Thus, we have the government of the United States also attempting in the late 1990s to exercise one of its favorite central powers of the twentieth century: monopoly-busting – in this case, against Microsoft and its alleged attempt to dictate the future of the Web, and therein of American and even world society.

MONOPOLY ONLINE?

The gist of the government's concern about Microsoft is this: "Microsoft is using the monopoly power enjoyed by its Windows operating system to fend off competition unfairly and extend its monopolistic reach," with the result that it "will gain a choke-hold on the Internet economy as people do everything from buying cars and airplane tickets to…reading news on the Internet" (Lohr, 1998a, p. D4). The specifics range from charges that Microsoft attempted to collusively split the market for Web browsers with its arch rival, Netscape, to claims that Microsoft has been pressuring manufacturers of computer hardware to make their equipment compatible only with Microsoft software (Vacco, 1998). But the general effect has been to make "everybody in the communications business...paranoid of Microsoft, including me," as Rupert Murdoch, himself no stranger to charges of monopoly in media, aptly put it (in Lohr, 1998b, p. D3). Thus, Jack Beatty, a senior editor of the venerable *Atlantic Monthly*, concluded his *New York Times* book review of Ron Chernow's *Titan: The Life of John D. Rockefeller, Sr.* with a lengthy paragraph describing the "emulation by Bill Gates [head of Microsoft] of Rockefeller's unparalleled instrument of competitive cruelty" (Beatty, 1998, p. 11).

Indeed, so mobilized has been the U.S. Federal government and its declining central powers in its attempt to check the "new Rockefeller" that it reached out to thirty state governments, which joined the Federal government in the suit against Microsoft. Such cooperation between national and state governments runs contrary to American political tradition – in which the national government has usually attempted to increase its powers over the states, which have always jealously guarded theirs – and is another consequence of heightening desperation over a shrinking centralized pie. With governments on all levels diminished by the decentralization of information, and unable to do much about such profound progress in media evolution, they might as well join forces against what they see as the single biggest centralized commercial source of this decentralization – Microsoft and its tilling of the Internet.

But is Microsoft indeed the monopoly invulnerable to all but governmental control that the government alleges it to be?

One would think, given the growth of decentralization, the erosion of centers, noticed as far back as the 1960s by McLuhan, that corporations would be no less subject to its sudden swings and vicissitudes than government – indeed, even more so, since laws (due to our representative form of democracy) are much less in tune with immediate public attitudes than are market preferences. And, in fact, that is exactly the case.

Microsoft began its meteoric rise in the early 1980s, with its introduction of MS-DOS as the operating system for all IBM personal computers and its soon-to-be legion of clones. Prior to this, CP/M had been the most popular operating system, animating early Kaypros, Osbornes, as well as several Apple and Commodore models (see McWilliams, 1982, for details). But CP/M was a problematic standard, since it was not compatible across such 8-bit machines – Kaypro's CP/M, though closely related to Osborne's, was not operational on an Osborne machine. When IBM began development of a 16-bit machine – which would become the IBM PC – it requested Digital Research, the corporate owner of CP/M, to provide a 16-bit version of that operating system. When Digital Research fell behind, Microsoft not only picked up the ball, but provided in MS-DOS a 16-bit operating system workable on all IBM and like machines. The growing community of computer users was delighted, and soon CP/M went the way of the daguerreotype and the silent movie.

By the end of the 1980s, the world of personal computer operating systems had been reduced to two: MS-DOS and Apple's Macintosh. Each system had certain distinct advantages. DOS gave its users better control over larger amounts of data – important for word processing, data management, and telecommunication, the three primary business uses of personal computers. But the icons and graphics of the Macintosh were more fun to use, and provided superior imagery for the burgeoning field of desktop publishing. Microsoft took the lead in developing an operating system that did the best of both, and today Windows is in more than 90 percent of personal computers.

Notice that in both of these cases, Microsoft did nothing collusive. Or, if it hypothetically did, that had nothing to do with its remarkable successes. First MS-DOS and then Windows triumphed for the obvious reason, often overlooked by Microsoft's detractors, that these operating systems did what the public wanted. At a time when a disk might have been hermetically sealed – so useless was it on any computer other than one made by the same company on which the disk was at first programmed or written upon – DOS made all IBM and clone-created disks interchangeable. At a slightly later time when the public yearned for operating systems with the power of DOS and the flair of the Mac, Windows emerged to provide just that service. In both instances, computer users at large, already empowered by increasing decentralization of information, called the shots.

Nor has Microsoft found a bed of roses in the public regarding software that the public did not prefer. The only reason that Microsoft had for allegedly attempting to collude with Netscape in the first place is that Netscape had jumped into such a commanding position in the Web browser market (completely obliterating the earlier, non-commercial Mosaic, which originated Web navigation as we now know it). And Microsoft's much heralded problems with Windows 95 show how difficult it is for a purported monopoly to force-feed even its most important new piece of software to the decentralized world.

Indeed, Microsoft is no more immune than the government to the potent centrifugal forces of the information revolution. The fact that Microsoft helped bring some of these into being gives it no privileged status – an instructive irony appearing often in history – for, as McLuhan and every media theorist worth his or her salt agree (e.g., see Postman, 1992, Meyrowitz, 1985), media have unintended consequences. Indeed, the very understanding of media begins with the recognition that the intentions of its inventors, and the expectations of its purveyors, are largely irrelevant to the use and impact of media in society. As I often point out, Edison first thought the phonograph he invented would make a great device for recording telephone conversations; within a decade, he came to see its main role as a photograph of musical performance; he then invented motion pictures (independently invented by the Lumières in France and Friese-Greene in England), and his first thought was that they would make a nice kind of accompaniment for the music on his recordings. Such thoughts, of course, were not absurd. A century after Edison had them, we have telephone answering machines and music videos. But the public, even in that far more centralized age at the end of the nineteenth century, had other thoughts, and thus held sway for a hundred years – or, until the public decided, again in complete independence of Edison, that it wanted answering machines and music videos (see Levinson, 1988b, 1997b, for more on invention and intention in media evolution).

This is not to say that the public is not influenced by media. To the contrary, the swelling tides of decentralization that crash against both government and industry are themselves the result, as we have seen, of media that provide centers of information everywhere. A more precise way of putting this complex dynamic would thus be: media make possible or further the expression of given human tendencies, with profound consequences for the social institutions of the time. Humans want to both lead, and be led (see Fromm, 1941, for discussion of the latter). The rise of electronic media in general, and digital personal computers in particular, has accentuated and focused the human penchant to lead – to hunt and gather information on our own, to make our own decisions, rather than be spoon-fed by central authority – as we walk into the new millennium.

But if central social institutions, whatever their remaining strength, are everywhere in jeopardy in the digital age, what of centers which are not only social but irreducibly physical in their extent?

We return, in the conclusion of this chapter, to its opening observation from McLuhan that "the city no longer exists." And we will see why McLuhan was in error on this point.

THE CITY ETERNAL: ANALOG AS DIGITAL

McLuhan of course was and is still right that a television program is the same whether watched on a screen in a home in a big city, a suburb, or a pit-stop out in the middle of nowhere. From the perspective of the TV viewer, there is no difference among the three – everywhere is somewhere, nowhere is no longer nowhere. Similarly, a fast-food restaurant, a chain hotel, a mall with outlets and stores either identical to or interchangeable with those of other malls, are all the same, whether located in or near a metropolis or a rest area on the turnpike. And in the years since McLuhan, online courses, online bookstores, online catalogs of libraries, with more to come, have dispersed to computer screens everywhere some of the informational bounties formerly unique to cities on the hill or in valleys, universities behind brick and ivy.

But is that all there is to life? Was the city nothing more than forms of information and commerce now readily reproducible en masse?

Certainly not.

It is said of New York City that one could eat in a different restaurant every night of one's life and still not scratch the surface. Nor would seeing the complete contents of the Museum of Modern Art or the Museum of Natural History online – when they do someday become therein available – be the same as seeing them in full dimension, even through glass or barriers, in the actual physical museum. Nor of course is exchanging the most scintillating, erotic letters online the same as making love. Nor is talking about taking a walk along a windswept beach, a forest trail, a city street bustling with people, or seeing pictures of such activities, remotely the same as doing them – as physically walking in those places.

This list of "nors" – infinite and ineluctable despite the digitally decentralizing tides we have been measuring in this chapter – stem both from the uniqueness of experiencing just about any aspect of life in the flesh, and the myriad of opportunities the city provides for such experiences.

The first has to do with our biological nature. We cannot live without information; in that sense it is crucial, with changes in its distribution via television and computers commensurately enormous in importance to us. Yet information is crucial only insofar as it keeps us in contact with – renders us reports about – a reality more primary. DNA is of interest as an information system because it fashions life, not other abstractions. Our senses of sight, sound, touch, smell, and taste enable us to successfully navigate the world they describe. Although sophisticated media such as television and the Internet may seem and indeed are often far removed from the real, day-to-day world we inhabit, the notion that these media are replacing that world gets punctured every time we turn from the TV or computer screen to answer the phone: the appeal of the telephone ringing is that there is a real person, likely someone we know, perhaps care about, at the other end of the line. Telephone usually trumps other media because its information is closer than theirs to our immediate reality. A knock on the door is thus even more commanding than the phone. Place thus persists as a bedrock of life despite the growing placelessness of information. Or, rather, since information was always placeless – its separability from place is what enables it to be communicated in the first place – the increase in its placelessness has no fundamental effect on the priority of place in our lives. The sharp decline of centrality is, from the perspective of our need to be connected to the physical world, a change in quantity not quality. Social institutions become decentralized; our day-to-day interpersonal interactions continue by and large as before. We may take an online course and therein avoid travel to a physical classroom, but we'll likely use the extra time to work in the garden or linger longer over dinner with a loved one.

This brings us to the second enduring advantage of the analog city in the digital age: it is not only a physical place, it is usually a plethora of places, from which we can choose. One of the benefits of personal computers and the Internet, as we have seen, is the great increase it gives us in choice of information, in contrast to broadcast media. Of course, years before either digital or broadcast media, we had cities. In bygone ages in which choices whether informational or physical were few, the city arose as their great maximizer – truly a computer in stone or brick or asphalt, in which people and their inclinations served as the software. Hence Samuel Johnson's astute observation that when a man tires of London, he tires of life.

Nowadays, a man or a woman can find vast arrays of information online to enjoy and to wield in business, politics, education, and the like. People can and do telecommute to work, or take online courses, from bayside cottages and mountain cabins far away from the city. Connected Education had students taking online courses from both kinds of sylvan abodes, and the administration of Connected Education – Tina and I – spent many summers in a cottage on Cape Cod Bay, seeing to Connect Ed business as well as swimming and playing on the beach with our children. That was undoubtedly a powerful exercising of choice, made possible by information technology, to live in one place and work or learn in another. And, yes, McLuhan was right that tourism is a big industry in many cities.

And yet, when the choices are tallied, we still find people moving into the cities, which are as vibrant as ever in the digital age. We find people opting for a maximization of physical choices – proximity to the central city – as they log on in increasing numbers to embrace the virtual, decentralized choices online. In this regard, Johnson remains more on point than McLuhan.

People, in other words, seem to want their cake of decentralization, and eat in the city too.

But we do not live by bread alone. There is more to life, and perforce communications and media, than the pressing issues of business, politics, and education we have considered in the past two chapters.

McLuhan also had a lot to say about aesthetics, and the impact of media upon them.

In the next chapter, we look at why stained glass windows in cathedrals of medieval cities, TV and computer screens in our present-day homes and offices, and blue skies everywhere are significantly different to behold than paintings, pages, and meadows. We look at McLuhan's distinction between "light through" and "light on" media, and the light it may shed on the new millennium.

8

THE MIND BEHIND THE SCREEN

"Most scholars use their knowledge as a flashlight," McLuhan remarked to me as we strolled around the trees near his home, squirrels in tow, at Wychwood Park in the summer of 1977 – "not to illuminate the world but to shine back into their own bedazzled eyes."

This worse than useless, self-deceptive light on the eyes – light coming from outside the eyes, not inside, shown back at the eyes via the externalization of interior knowledge in the flashlight – was but an extreme example of the "light-on" method McLuhan attempted to improve upon in his and our understanding of media. The "mosaic procedure, which I try to follow throughout, waits for light *through* the situation," McLuhan explained in 1960 (p. 11, italics in original). "It does not primarily try to play light *on* the situation."

But it was more than a procedure.

Whether it was coming from inside the eyes – not via flashlight shown back on the eyes – or beyond the skies, whether through a window from someplace outside or, eventually, through a television screen, McLuhan held "light-through" in media to be more profound and involving of the human observer than the more commonplace falling of light on the world and its things. Thus, for McLuhan, the contents of a room, illuminated by light shining on and off them, were far less interesting than the electric bulb through which the light was shining (another rendition of "the medium is the message"). Similarly, according to McLuhan, the reading of pages (unless they were illuminated manuscripts) and the viewing of movies are intrinsically less involving as processes than watching television – the audio-visual equivalent of the light-bulb, with light shooting through its screen rather than bouncing off its screen as in movie projection. The result is that, in television viewing, "illuminations project themselves at the viewer" (Carpenter & McLuhan, 1960, p. x) – the television becomes the projector and the viewer becomes the movie screen. No wonder TV is so involving.

Significantly, the personal computer too partakes of the light-through procedure.

Does this explain, in part, its astonishing colonization of pre-digital culture in the past few decades, with a speed rivalled only by television vis-à-vis radio and motion pictures in the 1950s?

Unlike the global village and centers everywhere and thus nowhere, intuitively obvious if not crystal clear in their application – or even the medium is the message, or acoustic space, not intuitively obvious at all, yet clearly pointing in the direction of something very important – McLuhan's notion of "light-on" versus "light-through" might seem the epitome of an interesting distinction blown well beyond its importance, a search for its expression and impact in media but a species of looking for the watch in the sky when someone says "time flies."

This uneasy mixture of metaphor and reality pervades all of McLuhan's work – it is part of its edge – but is especially prevalent in his dichotomous partitions of media, such as light-on versus light-through, and the much better known, and overlapping, hot versus cool, which we will consider in the next chapter.

Here, we will see that the "light" distinction has considerable value as a tool for explaining the emergence and impact of new media, as well as further illuminating – not in the flashlight mode, one hopes – some of the raw materials, the constituents, of McLuhan's approach to media that we first discussed in Chapter 2.

We begin with another look at McLuhan's approach, from a slightly different angle.

SCIENCE AND THE AESTHETICIAN

I recall reading years ago about a child of highly educated parents in the nineteenth century. They would see him, in the summer, carefully studying the flowers in their meadow, examining the texture and size of the petals, the shape and arrangement of the surrounding leaves, the placement of it all against the trees and the sky. The parents were sure their child was on his way to becoming a naturalist, a botanist in particular. They were wrong. The boy's intense interest in the colors and shapes and situation of the flowers in the field was an interest in colors, shapes, and arrangements in their own rights. He was on his way to becoming an art critic, not a scientist.

That child was not McLuhan – he was born in 1911 – but the boy's interest in the flowers, especially the proximity yet significant difference of that interest in relation to science, describes McLuhan's approach to media to a tee. A graduate student at Cambridge University in the 1930s – seat of the Cambridge Apostles and Russell and Moore at the turn of the century, and the literary criticism of I. A. Richards and the Bloomsbury group it in part gave rise to – McLuhan was accustomed to looking at the artifacts of the twentieth century, scientific and technological, as raw materials from which to construct an artistic edifice of understanding. Thus, although splitting the atom was obviously of extraordinary importance for practical weaponry as well as theoretical science, and McLuhan recognized it as such, he preferred to regard it as a better way of looking at the world – by shooting a stream of particles through it, rather than passively shining light upon it and studying it from afar. And indeed, although such understanding through perturbation was exactly what John Dewey and his pragmatic philosophy held as the best (and only reliably truth-generating) method of science, McLuhan staunchly resisted any appellation as a philosopher – "I don't explain – I explore" (McLuhan in Stearn, 1967, p. xii) – preferring instead to glean and provide insights from his probes as would a connoisseur of fine wines and bouquets. McLuhan was thus as distinct from Dewey as Richards from Moore – though Richards was literally Moore the philosopher's student – viewing philosophy, and especially its disputes over appropriate methods, as a distraction from the appreciation of the world and its technological mechanisms. Such appreciation was by no means approval, but a taking of the technological world as directly as possible, a strategy which McLuhan contended was most likely to reveal its hidden effects. In this sense – in his eschewal of formal philosophy and his treatment of science as an artifact – McLuhan's stance is both Baconian at the dawn of modern science (see the world as it is, not as you are taught to see it) and post-scientific (although he collects rather than tramples on science – otherwise I would not find his work of such interest).

One consequence of McLuhan's levelling of science and its technological applications to pieces in his mosaic is that the large and small discoveries, the big impacts and little twinges, often have equivalent weight in his schemas. Thus, the splitting of the atom is an in-depth way of looking through the world, a property shared by any window or screen: glass and atom-smashers alike give us a view of the inside, rather than what glances off the surface. The simple transparent properties of glass, in contrast to the reflective qualities of opaque surfaces, render this common technology – and the more sophisticated technologies like TV and computer screens that employ it – as profoundly informative in its own way as a particle chamber.

Of course, glass backed-up by silver serves as a mirror, and therein works as a deliberate exception to its otherwise light-through performance. McLuhan seizes on this technological commonplace too, and finds meaning in it for our understanding (or misunderstanding) of media: "The youth Narcissus mistook his own reflection in the water for another person. This extension of himself by mirror numbed his perceptions" (McLuhan, 1964, p. 51), with the result that he drowned. Looking at the surface of the water, rather than through it, thus proved not only superficial and misleading but deadly in this mythical instance.

Not that looking through things is easy or without risk. But it does provide us with an opportunity to get beyond our own reflections.

A KINETIC UNDERSTANDING OF MEDIA

The first photographs – developed by L. J. M. Daguerre in the 1830s, and known, appropriately enough, as "daguerreotypes" – are quite literally the high-point of light-on technology in media. Wrought as immediate positives on silvery surfaces, they served as mirrors of their beholders when tilted at one angle, and as reliable, eternal images of their subjects when tilted to another. The mirror effect was unintentional – i.e., the silvered surface was the only way Daguerre could get his images to endure – but who knows how many owners of a small, pocket daguerreotype stole a quick glance at their own fleeting reflection when they were not enthralled in the miraculous permanent image of their loved one (which perhaps in some egotistical cases was also an image of the beholder). In any case, Fox Talbot in England and others soon figured out how to make images endure without the silvered mirror surface – indeed, how to make negatives, each of which could be used to make many positive copies, on a variety of media including paper. None of these positives were literal mirrors, and thus the still photograph subsided to the basic level of light-on media that characterizes pages of print, paintings, and indeed most things we regard in this world.

And yet the fixing of an image for all time on a photographic plate, mirrored or not, provoked a kinetic response that would revolutionize art, photography, and eventually produce the visual media of television and the personal computer.

Art – painting in particular – was the first to register the effect. Contrary to painter Paul Delaroche's premature exclamation, upon his initial sighting of a daguerreotype, that "From today on, painting is dead!," the venerable hand-crafted medium went on to thrive by doing something other and more than the still photograph's remarkable stopping of the world in motion. Instead, Impressionism deliberately attempted to suggest the transitory image, tried to capture and convey the passage of light itself by daring use of broken color and sketchy strokes of the paintbrush. By the time of Cézanne and Seurat at the end of the nineteenth century, the successors of Impressionism had gone from the illusion of capturing to producing light on their canvas. McLuhan (McLuhan & Parker, 1968, pp. 24–5, italics in original) remarks of this greater illusion: "The Impressionists painted light *on*. Seurat painted light *through*, making painting itself the light source, and anticipating Rouault's recovery of the stained-glass-window effect of light through."

Meanwhile, photography proceeded to pursue motion by more literal means. If, as André Bazin (1967, pp. 14, 15) aptly observed, the still photograph rescues an image from its "proper corruption" in time, motion photography reinserts that image into a sequence of time, and conveys "change mummified, as it were." Although the motion in movies is in one sense an illusion – conveyed by a series of still images played so quickly to the eye that it sees the images as a continuously moving piece – there is real motion on both sides of the divide, in the reality initially captured by the series of still photographs, and in the motion of the film strip through the projector (earlier, the circular motion of the plates in Edison's kinetoscope). Motion photography is thus a classic example of what I call the "instant coffee" of communication (actually, a more mundane rendition of Shannon–Weaver's encoding/decoding): the motion of the world (coffee) is converted into a series of still images, so as to be suitable for transport (coffee powder), which in turn are reconverted into motion as the film moves through the projector (hot water is poured on to the powder). Unlike the Impressionistic suggestion of light in transit, the motion picture actually operates via light in transit – through the projector – with the result that the pure suggestion of the painting is replaced by a mixture of suggestion and reality in the movie.

Nonetheless, Bazin's invocation of "change mummified" highlights both a distance that the film viewer maintains from the screen, and the impossibility of changing any of the action within the movie, on the screen, once the show is underway. The movie for all its motion is still a projection on to a surface, and thus identified by McLuhan as light-on.

But how can the motion picture, which at least entails the real motion of projection, be light-on, whereas the pointillistic post-Impressionism of Seurat, which is after all in reality just static dabs of color on a canvas, be identified by McLuhan as light-through?

Half of the answer lies in the deliberate incompleteness of pointillism, whose points of pure color (i.e., light) are supposed to be integrated by the viewer into vivid effects. The motion picture experience, in contrast, at least appears on the surface to be a more complete package – we are not aware of the constituent individual images when we see them all in motion. Thus, part of the kinetic impact of light-through media – much like the related impact of "cool" media (a much more celebrated McLuhanesque appellation) that we will consider at length in the next chapter – comes from the active participation they invoke.

The other half of the answer points to a fundamental characteristic of McLuhan's approach to media – especially in his either/or distinctions of light-on/light-through and hot/cool – which, if not appreciated, can lead to much confusion and infuriation. This characteristic is: the distinctions work best, make the most amount of at least immediate, obvious sense, when they are applied to media that attempt to do the same thing, or appeal to the same modes of human perception. Thus, pointillism as a light-through medium can best be appreciated not in comparison to motion pictures, but in comparison to other forms of painting – to wit, earlier Impressionism, and, even better, portraiture and landscape art, which were as "light-on" and complete as the still photograph.

The medium that McLuhan had in mind as an apt light-through audio-visual apposite to light-on motion pictures was television.

THE PERIPATETIC COUCH POTATO

Only McLuhan could come up with a dichotomy which held that the motion picture, which for most of the twentieth century most people went out of their homes to see, was less involving, less active in the relationship it evoked from its viewers than television, which most people sprawl in front of in various states of consciousness, on couches and easy chairs and beds in their homes, to see.

Upon closer inspection, however, McLuhan's logic becomes clear. The motion picture is comprised of frames which display a photochemical registration of a reality that once was – at its core, the motion picture is static, or a bringing back into motion of a series of frames which are themselves utterly unmoving. In contrast, although television can and usually does present a motion picture, the process of television is an electronic transmission in the present. In the case of a live presentation, it is also a transmission *of* the present, allowing us to see events unfold as they happen, newscasts subject at any time to change, "breaking" developments in a way that the motion picture is constitutionally incapable of presenting. For that reason alone, watching television is a more active, involving experience, in which the viewer at all times is responding, or ready to respond, to actual ongoing events. That the viewer may be slouching ready to respond on a chair, rather than standing ready to respond in a pressed suit of clothes, makes the readiness no less profound.

McLuhan (1964, pp. 272–3, italics in original) puts the matter as follows: "With TV, the viewer is the screen. He is bombarded with light impulses.... The TV image is visually low in data. The TV image is not a *still* shot. It is not a photo in any sense, but a ceaselessly forming contour of things limned by the scanning finger. The resulting plastic contour appears by light *through*, not light *on*..."

Characteristically, McLuhan is making a variety of overlapping yet distinct observations here. The active, living quality of the TV image – "ceaselessly forming contour" – and its departure from the static photographic image is one. The idea that the TV image is involving because it "is visually low in data" is another, and the heart of McLuhan's hot/cool distinction, which we will consider in the next chapter. "With TV, the viewer is the screen" is yet a third. It is perhaps the most provocative, and the heart of the light-on/light-through taxonomy of media.

For in reversing the movie theater set-up in which the projection begins behind the viewer and ends up on a screen in front, to the TV set in which the projection begins behind a screen in front of the viewer and winds up literally in/on the viewer's face, McLuhan is giving us a clue as to why light-through is irresistible. By originating beyond rather than behind us, by suggesting something more rather than what already was, light-through is inevitably a finger beckoning us to further investigate.

In the case of television, as we have seen earlier in this book, there were and are still inherent limitations in our capacity to investigate and influence the presentation of a mass medium.

These by and large disappear when the screen that the light comes through belongs to a computer rather than a television.

THE SCREENS AND THE SKIES

Even the light-through computer screen is not entirely free of the narcissistic mirror: certainly the writer who looks at his or her words on the screen is regarding a daguerreotype of sorts of the late twentieth and ensuing century. But once connected to the Web, the personal computer metamorphoses from word processor to word selector – and, increasingly, image and sound selectors as well – and thus the screen becomes a portal to a virtual infinity (in both senses of the word "virtual") of possibilities beyond.

The analogy of these distant points on the screen to the stars in the sky is striking. The sun, of course, is also light-through the sky, but so powerful is its energy that all we get from it – other than light-through blueness of sky and tufts of clouds – is its light-on illumination of things on Earth. In contrast, the stars beckon us beyond. (The moon and the planets of our solar system, though light-on in terms of receiving illumination from the sun rather then generating it themselves, nonetheless seem light-through and mysterious to us, since they reflect light to us from the same other side of sky as the sun and the stars, with a quality much like starlight.)

The difference between the sun and the stars, between the words in books and on computer screens, is more than aesthetic. As we saw in the previous chapter, the sun is a beacon of centralization, whereas the stars, in Bruno's time as in our own, are harbingers of a profoundly decentralized humanity. Similarly the book, a radically decentralizing medium in comparison to the handwritten manuscript, serves as a reliable, enduring, "center" of the information on its pages in comparison to the continuous change of letters of the computer screen (we see, again, the relative nature of characterizations in media). Indeed, our continuing need for centers –

for knowing that the words we expect will be on a given page whenever we look at it – is likely the only reason that books will continue as an important medium in the twenty-first century and after. (Of course, we might deliberately design a computer screen so that the words it displayed never changed. But then it would just be a special kind of book; see Levinson, 1998a, for more.)

That McLuhan seized on the aesthetics of looking out of windows and related this to watching television but not movies, that this aesthetic applies both to computer screens that barely existed in McLuhan's day and to starry skies that have existed for all of humanity, and that this aesthetic has implications for numerous aspects of life that go well beyond aesthetics – all of these facets of light-on/light-through capture the brilliance of McLuhan's thinking, its encyclopedic reach, and its continuing value to us. Over and over again, we find him almost obsessing on a seemingly minor distinction – sometimes a distinction not only minor but difficult to see – and blowing it up out of all apparent proportion. And then we're suddenly hit in the face with how relevant it is to an event, or an impact of a medium, unfolding right before our very eyes.

Today, light not only flows to us through our computer from a myriad of different hypertexted places, but we reach back into the computer, and out to real and virtual worlds, to buy things, do business, make friends, pursue relationships.

And we look to the stars. For just as the computer and its screen has moved our information just about anywhere we want it on Earth, and indeed made our physical transport easier too (we can buy cars and airline tickets on the Internet), so the screen of the sky leads us to visit the moon, and the planets, and someday the stars beyond.

The common ingredient in both light-through experiences is something beyond, something we do not already fully know, in a word: the future. The ultimate light-through.

Closely related to the uncertain, intriguing, beckoning future of light-through is a quality of media McLuhan called "coolness." He held the motion picture, brightly bouncing off the big screen, to be "hot," in contrast to the restless, blurry image of the TV, which was "cool." Much as in light-on versus light-through, the first washes over us, splashes us, shocks us, whereas the second draws us into its world.

In the next chapter, we examine and update these and many other characteristics and consequences of "hot" and "cool" media – one of McLuhan's most vivid, and criticized, dichotomies.

9

WAY COOL TEXT

McLuhan's distinction between hot and cool is one of his most celebrated, misunderstood, and useful tools for understanding the impact of new media. Gerald Stearn's 1967 anthology of often sharply critical commentary on McLuhan's work – with contributions by such notables as Tom Wolfe, George Steiner, and Susan Sontag – is aptly titled *McLuhan: Hot and Cool*. Haskell Wexler named his acclaimed 1969 movie about media and politics *Medium Cool*. In *Annie Hall* (1973), Woody Allen waiting in a movie theater line pulls McLuhan out of the shadows to correct a professor pompously holding forth that television is "hot": "You know nothing of my work," McLuhan helpfully offers, "you mean my whole fallacy is wrong." The professor was perhaps misinformed by Jonathan Miller's *Marshall McLuhan*, the 1971 "Modern Masters" book, which blithely states that "the term 'hot,' then, applies to those messages that have gaps in their information structure, requiring an act of positive inference from the recipient" (p. 97) – or precisely the opposite of McLuhan's usage.

Like most of McLuhan's probes, the hot/cool thermometer can provide an immediate rush of presumed comprehension, followed by long periods of frustrating, even maddening, attempts to fit together readings that apparently defy integration. Such frustration, alas, has been the last stop for many academic and casual readers of McLuhan, who fail to reap the rewards that a sustained and fully reasoned pursuit of hot and cool can provide. Chief among these benefits are assessments and predictions for our wider culture – weather reports, to invoke the ecological analogy favored by McLuhan.

Of course, one often gets an initial sense of the temperature by just putting a finger up to the wind. The result is far from immediately clear or well organized – and this is the case for McLuhan's taking the temperature of media – but it may put us in better touch with what is actually going on in the world than a host of statistical surveys and abstract ideologies.

McLuhan's invocation of hot and cool derived from jazz slang for brassy, big band music that overpowers and intoxicates the soul (hot) versus wispy, tinkly sketches of sound that intrigue and seduce the psyche (cool). The brassiness of the big hot band bounces off us, knocks us out – we neither embrace it nor are immersed in it – in contrast to the cool tones that breeze through us and bid our senses to follow like the Pied Piper.

McLuhan's first full-blown presentation of this dichotomy and its application to communications media appears in *Understanding Media* (1964), which builds upon his discussion of HD (high definition) versus LD (low definition) media in his 1960 typescript "Report on Project in Understanding New Media." His idea was that hot media have loud, bright, high-profile deliveries of information that accost and sweep over our soon satiated senses, in contrast to cool media whose blurry, soft, low-profile demeanors invite our involvement to complete or bring to life the quiet evening. The now classic players include hot, big screen, technicolor motion picture theaters versus cool, small screen, black-and-white TV; hot printed prose in novels and newspapers versus cool poetry and graffiti; true-to-life photography versus the spare suggestive lines of political cartoons; and the fuller dimensions of sound and music on radio, hi-fi ("high-fidelity"), and stereo versus the irresistible tin ear of telephone.

Further, and more importantly, McLuhan saw such hot coals and cool winds regulating the temperature of the cultural environments they inform and in turn draw energy from: The 1930s, running on radio and motion pictures, were a hot age. Vivid colors, sharp hairdos, keen wit and articulation were in. By the 1960s, television had cooled down the culture to the point that worn jeans, unkempt hair, and an inchoate getting in touch with one's feelings were de rigueur for the leading edge of style.

But on further reflection, how exactly can television, which conveys both images and sounds, be cool in comparison to radio and its sound-only? And whereas prose in a textbook may be less participational than poetry, surely prose in a love letter may provoke the most involvement of all.

The latter example touches the heart of this chapter: Online communication – e-mail, group discussions, digital text in cyberspace – is by all standards the most fully interactive medium in history, and much more ephemeral, sketchy, wide-ranging, fast-moving than print fixed on any paper. Online text thus seems cool to the point of approaching Kelvin's zero. Yet it traffics patently in a species of print, and mostly prose at that – media whose aspects when expressed in newspapers, magazines, and books are among McLuhan's favorite examples of hot, high-definition packages.

Does this mean that McLuhan's hot-and-cool taxonomy is broken, stretched beyond serviceability and rendered undeniable mind junk at last on the shoals of the digital age?

In this chapter, we'll see why this is certainly not the case – we'll see, indeed, how and why text rendered electronically inevitably acquires and radiates a coolness that far exceeds that of television, and is in synch with the success of such cool recent forms as rap music and Quentin Tarantino movies.

To understand this process, we first have to more carefully examine what the hot/cool distinction means – how it in fact operates – and what constitutes a medium, or agent of a hot/cool impact.

BARE ESSENTIALS OF THE MEDIA THERMOMETER

McLuhan thinks television with its sounds and pictures is cool, less complete, in comparison to imageless radio, and for that matter to most words printed on a page lacking sound as well as image.

How can that be?

Because the coolness of a medium, its invitation to fill in the details, comes not from the number of senses it engages, but from the degree of intensity of its engagements. The sound on most television today has far less clarity, depth, and range than the output of radio, "walkmen," and CDs. This flows from television's origin as a talking-head medium, in contrast to radio (at least after Sarnoff in 1915; see Dunlap, 1951, p. 56), tape players, and CDs as conveyors of music. And the images on TV – even years after television grew colors, trebled its original size, and developed a memory via VCR – are still less overwhelming than the pictures in movie theaters, and less specifically retrievable, more fleeting, than words printed on the pages of a book.

Thus television began and remains as a cool medium, despite its cutting a wider sensory swath than some of its rivals. We might say that the expanse of a medium – whether gauged by the number of senses it appeals to, or the number of people it reaches – works as a multiplier or accelerant of its coolness (or hotness), provided that the components of the expanse all have an equivalent media temperature. More skin, less cover, is generally more cool, more inviting.

Of course, comparisons of media temperature, like comparisons of any individual factor, are easier to ascertain to the extent that other factors are the same. Television is most obviously cool in comparison to the movie theater, and its equivalent audio-visual array. The postage-stamp speaker in the telephone's earpiece most clearly sets off its coolness in comparison to radio's heat (though, as we will see below, the inherent interactivity of the telephone also figures prominently in its coolness), just as the brevity of poetry trumps most prose in coolness in an immediately apparent way.

Such clear-cut comparisons are useful starting points in any exploration, elaboration, and application of the hot/cool taxonomy of media. The key is not to get stuck in them – to be ready to investigate the possibility, as we just did above, that television is cool not only in comparison to its audio-visual sibling the movie theater, but in contrast to radio and print as well. Hotness and coolness, in other words, are more than relative measurements of one medium to another: they are rather properties of a medium's usage by humans, properties that to some degree persist regardless of the presence of one or another medium in the environment.

But neither should we conclude from this that television or any given medium has some sort of eternal, unchanging, metaphysical claim on coolness (or hotness). To the contrary, media constantly undergo evolution under pressure of human usage and invention, and this can at any time register a profound change of temperature. Television, as indicated above, has become hotter since its introduction in the late 1940s, now offering color, two-foot or larger screens, and retrievability via VCRs. That it continues to function as a cool medium means that such changes, these "hotting ups," to use McLuhan's parlance, have not been hot enough to burn through and transform TV's cool packaging, i.e., the incompleteness of presentation that one still receives from watching a full-size, color SONY with a VCR attachment in one's living room.

Similarly, the printed word cooled down considerably in the twentieth century, under the summer breeze of paperbacks, McLuhan's books themselves, and deliberate attempts to air-condition the pages and format of magazines such as *WIRED*. But as unlike the nineteenth-century tome and periodical as today's sleekest, multi-dimensional media of print may be, they nonetheless share the fixation of words to paper that has been the defining characteristic of writing since its inception, and renders it hot in comparison to more fleeting media ranging from speech to television.

The advent of writing on screens, however – our cross-breeding of a new medium that seems an almost 50/50 hybrid of hot text and cool TV – is a different matter. Will the traits only cancel each other out, leaving life on the computer screen neither hot nor cool, just lukewarm lake water, remarkable for neither its high profile nor its invitation? We already know that is not the case. But then, which trait, which temperature, will ultimately be dominant? Or will the result be a medium with patches of each?

The answer hinges on how media perform as the content of other media. As we saw in Chapter 3, every medium is like a Chinese box or a nesting doll – a medium within a medium within a medium, going back to thought itself – and when we experience any given medium, we hear the voices, see the faces, feel the breath, of all the media that have come before. What makes each new medium different, then, is the specific way it commends to our attention the prior media that it may encompass. When shown in a movie theater, film is a hot, bigger-than-life, theatrical experience, complete with laughing, gasping, sobbing, and applauding audiences. When shown on TV, the same film becomes content for a very different medium, at once icier, isolational, and more in need of the warmth of our participation. And when shown via a VCR, with the capacity it gives us for stopping and contemplating, for replaying, for moving quickly ahead, that same film becomes part of a significantly different medium yet again – it becomes, in Nikita Khrushchev's remarkably apt description upon seeing a video tape of his "Kitchen" debate with Richard Nixon in July 1959, "like reading" (see MSNBC-TV, 1997).

This suggests that, however hot the legacy of the printed word, it will yield to the cooler currents of words in motion on screens, since those infinitely refreshable and thus indefinite screens are the proximate media that deliver the words.

But this triumph of cool entails a factor that goes beyond the properties of the personal computer and its Web applications. Not only do hot and cool not operate in a vacuum for individual media – not only are these media characteristics best observed in comparison of one like medium to another – but they also have an effect on the overall culture, which in turn selects media of appropriate temperature.

Indeed, for McLuhan, the most important result of the rise of television as a cool medium is that it ventilated our culture, our attitudes, our interpersonal behavior, leading us to prize the whispered invitation to the stuffy announcement.

Elements of hot culture which dominated the first half of the twentieth century of course continued to have impact, but the general lowering of temperature in the age of television made the advent of cool computers no coincidence. Our culture was waiting for them.

CULTURAL CONSEQUENCES

The terms hot and cool were already laden with connotation well beyond their jazz origins when McLuhan first started applying them to media in 1964. And their popularity has continued to expand.

To be cool has long been desirable – not only that, but "way cool" for at least the past decade. But hot is no small potatoes either. Certainly, at least one automobile company thinks so – "Toyota's hot, hot, hot!"

Cool connotes a profound, effortless synchronicity with the universe as it actually is and will likely be, speaks softly of deep pools, of being in tune with the future. Hot is fast cars indeed, fast food, life in the fast lane, encounters quick, overwhelming, intense – hot buns, hot abs, hot babes and hunks, embrassez-moi, run me over and leave me senseless.

The cultural ideal, apparently, is to be either hot or cool in the appropriate environment, i.e., in situations or eras that most reward or require whichever.

This is often not easy. In his two pertinent chapters in *Understanding Media* ("Media Hot and Cold," "Reversal of the Overheated Medium"), McLuhan not only indicates how the temperature of a medium or constellation of media can set the heat of the larger culture, but how hot cultures often provoke the emergence of cooling media, and vice versa, in a thermostatic or balancing function. McLuhan developed this notion from his fellow Canadian and mentor-in-media Harold Innis' view that cultures do a see-saw act between time-binding or preservational media and space-extending or disseminative media – see Innis' *Empire and Communications*, 1950, and *The Bias of Communication*, 1951. McLuhan later expanded this view into his "laws of media" or "tetrad" theory (discussed in Chapter 15 of this book). Given such flux, one can often get caught with one's pants down – or at very least, dressed inappropriately for the media occasion.

Peter Cook and Dudley Moore's classic *Bedazzled* (1967) portrays this problem beautifully. Moore strikes a Faustian deal with the Devil to make seven wishes come true. He asks in one of them to become an Elvis-like rock star. Alas, his joy is short-lived, for by the time this movie takes place, hot Elvis has been replaced by cool Mick Jagger as the rock icon. Poor Moore/Elvis is out of synch with the temperature of the times. (And Jagger continued as an icon of coolness in the 1990s – at least for baby boomers like Bill Gates – as evidenced by Microsoft's use of the Stones' "Start Me Up" as theme music for its Windows 95 TV commercials.)

McLuhan had a field day plumbing for hot-and-cool vectors in American politics. Adlai Stevenson's literacy was an early victim of the television age, which favored the cipher that was Eisenhower. But JFK's wit and poetry played perfectly to the camera: surveys conducted after the Kennedy–Nixon debates showed Kennedy won the debate on television, while Nixon carried the day in the same debate heard on hot, serious, articulate radio. Eight years later, however, Hubert Humphrey's in-your-face ebullience made him a much hotter candidate than Nixon, who had learned to somewhat disguise his face and his logic from the camera. But TV had the last laugh on Nixon in Watergate, in which his earnest denials of guilt, reasonable enough in print and voice, came across as a cartoon villain's shifty-eyed protestations on the tube. A classic political cartoon by Herblock from that era pictured Nixon crouched against the side of an overturned desk in the White House, ammunition ready, shouting "Come and get me, copper!"

Most of these hot-and-cool interpretations have a degree of plausibility, and the examples ring true. Yet surely other factors played a role in the rise and fall of those political fortunes. McLuhan, in any case, did not live to see the strongest challenge to hot-and-cool exegesis of politics in the 1980s, when Ronald Reagan, a hotter candidate than the inarticulate Jimmy Carter and low-key Walter Mondale by most standards, outpolled his rivals by big margins in two elections.

The success of Reagan's old-time warmth in a cool age could be explained by recourse to hot elements in the 1980s – say, the heating up of television described above, or even the temperature of early personal computers, much more dependent upon fixed, rigid, i.e., hot, programs than the more open architecture that followed. One could also seek elements of coolness in Reagan's style – his trademarked capacity for saying little or nothing so eloquently, honed by his work in television commercials in the 1950s.

But the better lesson for hot/cool cultural assessment to be drawn from the Reagan years is that hot and cool as a tool of analysis has its limits when applied to politics. Indeed, we might construct a general principle from this: the further away from its media origins, the more tentative, conflicted, attenuated the hot/cool analysis becomes. Applied to music, movies, and clothes (which, as McLuhan noted, are in many cases more a medium of personal communication than a means of protection from the elements), it works fine.

It has its innings in politics. Although George Bush's victory over the practically tongue-tied Michael Dukakis in 1988 was another hot success, Clinton in most respects was certainly cooler than Bush in 1992 and Bob Dole in 1996.

As an example of where hot and cool eddies in overall culture have least impact, we might consider the enterprise of science. Although cultural expectations no doubt direct scientific research and govern initial interpretations of findings, the evidence of external reality sooner or later cuts through prevalent beliefs – otherwise, we would not have scientific revolutions on the level of Copernicus, Darwin, and Einstein. The key factor here is an external reality which, although filtered to us via our perceptions, beliefs, and media, exists to a large extent independent of them. Fritjof Capra may have given *The Tao of Physics* (1975) a popular spin – a song that continues playing today in books like Tipler's *The Physics of Immortality* (1994), which claims the future already exists and governs our present – but the real progress in science still comes from the hard edges of experiment and observation. Reality bites, whether the temperature is hot or cool.

That our popular culture has undergone a pervasive cooling since the 1950s, however, there can be no doubt. Soft colors, soft voices, software – all offer quiet basins of attraction that pull forth our participation. A signal irony here is that television, which led the way to the shady side of the street, is obviously not intrinsically an interactive medium. McLuhan saw this lack of interactivity as the source of one of television's most important effects: stimulated to participate by a medium that does not permit it, viewers develop an insatiable need to reach out and touch someone, to get in touch – to reach through the light-through medium – and to be in as much literal physical contact with others as possible. Everything from touchy-feely groups to the Beatles' "I Wanna Hold Your Hand" to the sexual revolution follows or at least gets impetus from this.

But what happens when a medium issues a soft invitation to interaction that *can* be accommodated – more, pursued as never before – via the medium itself? What happens when the participational possibilities are hardwired into the very system that tempts us with its low-profile, incomplete presentation?

We get electronic text and its consequences.

TEXT INTERACTIVE

Although television received most of the credit for the cool age, it was not the first low-profile medium in the communications revolution, and perhaps not even the most important. That priority goes to the telephone, an 1876 invention that was child and perhaps partial catalyst of a cool age of Impressionism and its progeny in painting, poetry, and music.

The relatively poor quality of telephone sound – it delivers but a gloss of the full human voice – made it from the outset a quintessentially cool medium. But it had a far more pressing reason to elicit interaction: namely, that the telephone is essentially, regardless of the quality of its sound, an interactive instrument. Indeed, it is a medium whose ringing can barely be gainsaid. The hope, however slim and unrealistic, that the caller will at last be providing the means for fulfilling my career, my fantasy, my need to be recognized by everyone or someone in the world, is almost always irresistible. Few if any other activities can take precedence. Communication theorists have long joked about couples afflicted by "telephonus interruptus" in their love-making, and the joke is often but truth.

So powerful, then, is the interactive pull, the tug of a live human being on the other end, that telephone would have been a cool medium regardless of the intensity and clarity of its sound. Indeed, we might speculate that telephone remained cool – the quality of its sound has not profoundly increased in the past hundred years – because it was locked into a keenly satisfying cool loop by virtue of its interactivity. Rather than a low profile that engendered interactivity, the interactivity of the telephone sealed its low profile, made any kind of higher profile unnecessary.

In contrast, non-interactive cool television, as discussed above, rapidly evolved to some degree of heat in its colors, size, and retrievability. This may also account for the enormous influence of TV upon our culture, in contrast to the telephone. Television, for all its ephemerality, was something to see, to focus on. Telephone deals in evanescent information – the answering machine only somewhat counters this – and as such it may have been too cool, may have operated too pervasively and unconsciously in our culture, to exert major discernible influence upon it.

But what consequences ensue when visual text – prose, an archetypal hot medium – becomes the content of cool telephone?

The early take on personal computers – early but still prevalent among media critics such as Neil Postman (for example, in his speech at the New School for Social Research in honor of John Culkin – to whom this book is dedicated – in February 1994) – was that they were some kind of souped-up television screen, likely to have similar impact on the cognitive process (namely: either none or deleterious). More enlightened users – and I immodestly include myself among them – thought that computers could be more appropriately considered new kinds of books. (As we saw in Chapter 7, the Supreme Court apparently agreed, striking down the Communications Decency Act in 1997 on the grounds that the Internet is more like a print medium than a broadcast medium, and therefore entitled to fuller First Amendment protections; see Greenhouse, 1997.) In any event, when connected to the world-wide telephone system, the personal computer immediately transcends both TV and the book: it becomes a special type of telephone, not only retaining its powerful cool interactivity but amplifying it. It actually has not two but three parents – books, telephone, and television – and the last two are cool. Online text thus comes by its coolness even better than naturally.

The onlining of text – the immersion of hot print in the cool interactive milieu of the global telephone network, outfitted with screens – has been a baptismal breakpoint in the history of writing, and perforce of our human existence. The fixed, linear specificity of first alphabetic writing and then print had long been one of its defining characteristics, rendering it a hot medium par excellence. Socrates was an early complainer about this heat, worrying in the *Phaedrus* that the written transcript, unlike verbal dialogue, provides but one unvarying answer to its questioners. He yearned for "an intelligent writing which…can defend itself, and knows when to speak and when to be silent" (Plato's *Phaedrus*, secs 275–6). Of course, living authors of even the most hidebound texts could always in principle be questioned as to their meanings, and text in the form of personal letters has always been intrinsically interactive, a sort of highly delayed, asynchronous telephone conversation, looked at in retrospect. But time is inextricably of the essence in human affairs, and the degree of delay in written questioning of living authors and personal letter writing – the time usually consumed for response and re-response in such asynchronous interactions – can deprive this interactivity of much of its impact. Perhaps this is why I. A. Richards wisely counseled in his *Practical Criticism* (1929) to let texts speak for themselves, insofar as this was possible, since an author questioned years after the act of writing might well be no longer reliable as guide to what was intended in that act.

Text rendered online, however, has the capacity for instant interactivity. Indeed, as early online users of American bulletin boards and commercial systems and the French "Minitel" discovered by the mid-1980s, text as a vehicle of synchronous immediate conversation could be more provocative than verbal exchanges. McLuhan's concept of cool can help us understand why: Conversations on the phone come in packages that usually tell the conversing parties who they are, their general emotional state from tone of voice, etc. These details can be seen as heating elements carried over from in-person conversations, in which an enormous amount of hot, high-profile detail is available: their retention on the phone serves to heat up, or slightly "de-ice," the telephone environment. But they are missing entirely in an online text community, where participants in a conversation have only each other's written words to go and know each other by. No wonder that inhibitions are down when fingers do the limbo on keyboards and mice in the middle of the night; no wonder that total strangers are propositioned and relationships virtually consummated in online chats now going on the world over. Text on telephone lines is even cooler, more seductive, than speech – it is often addictive (in a psychological sense) precisely because its mode of presentation prevents us from ever getting enough of it.

A dollop of asynchronicity can heighten the impact of this potent mix. Conversationists on Usenet lists see their comments responded to in minutes, hours, almost never longer than a few days. The Western Behavioral Sciences Institute in the early 1980s found such a tempo ideal for intellectual dialog: the combination of the more time allowed for considered response, the endurance of the record, and the global situation of participants made online seminars far more productive than the typical in-person meeting. Online courses since conducted for academic credit have recorded similar, consistent results: asynchronous interactivity, when paced in minutes and hours and days online rather than days and weeks and months offline, engenders a level of participation in quality and quantity that often exceeds that of the typical in-person classroom, in which only one person (usually the teacher) can speak at any one time (see, Levinson, 1995b, for details). A cool middle ground is achieved, one that maximizes participation because it is free of two sets of inhibiting factors – those that characterize the in-your-face-in-person meeting, and those attendant to the molasses interactivity of traditional street-mail letter-writing and the non-interactivity of taking books out of a library. Online learning thus improves upon place-based, book-paced (which also could be spelled book-paste) education. It conveys some of the benefits of the older tutorial system that mass education had itself largely supplanted (except in tutorial systems still in operation in Oxford and Cambridge, and more recently established in the British Open University system) – with the key difference that online education is in principle global in reach, in contrast to the inevitable locality of in-person education whether on the tutorial or mass level.

McLuhan would no doubt have been an advocate of cool online education. "My entire syllabus," he writes in his "Report on Understanding New Media" (1960, pp. 14–15), "proceeds on the principle that low definition (LD) is necessary to good teaching. The completely expressed package offers no opportunity for student participation. . . . LD media like telephone and television are major education instruments because they offer inadequate information." Although his expectations have been fulfilled for television only partly and for voice telephone not at all, the online course conducted via personal computers that send and receive text via telephones connected to "computer conferencing" facilities on central computers and the Internet has become an ideal forum for cool "good teaching" – as defined not only by McLuhan, but education theorists ranging from Dewey to Piaget to Montessori, who have all emphasized the importance of the teacher as facilitator (not lecturer) and the active student learner (see Perkinson, 1982, for more on the philosophy and history of non-passive education).

Unsurprisingly, the cool impetus of text interactivity has picked up energy and mass as it has evolved. The evolution of hypertext and its links – predicted decades ago, albeit on centralized mainframe systems that never quite came to be, by Ted Nelson (see, Nelson, 1980/1990 for a summary of his 1960s work) – offers the online user a myriad of possible hidden knowledge connections. You could click on McLuhan, were you reading this document in online hypertext, and who knows where that might lead. You would likely be linked to online repositories of McLuhan's work, to the extent they exist, but you would also likely have access to work relevant to McLuhan that does not yet even exist at the time of this writing. On the one hand, you could rest assured that a huge amount of knowledge is little more than a keystroke or two away; on the other hand, neither you nor I nor anyone can ever know at any time just what the extent of that knowledge is, because it is constantly being rearranged, added to, linked to new links ad infinitum in possibility. The Web and its hyperlinks thus comprise a quintessential case of a cool system – a verdant breeze wending its way through every leaf in the hothouse of knowledge, not only cooling but pollinating as it moves along.

No wonder that the Web and its pages are catnip to anyone with a taste for intellectual inquiry – and no wonder too, as we will see in the next chapter, that it is dramatically transforming the very nature of publishing. Meanwhile, the deeply cool procedure of every reader creating his or her own book, different with each navigation of the hypertext Internet, has already connected with other cooling aspects of our popular culture. What is *Pulp Fiction* (1994) (and *Reservoir Dogs* before it) if not a hypertext movie, a presentation of disparate though intricately related swatches of narrative, with the viewer invited to put them together to tell his or her story? Rap music and its minimal and open structure offers similar invitations. In both cases, we see hot media – like text before the Internet – cooling down under the influence of a major plunge in temperature in our cultural environment. (See Benzon, 1993, for some interesting speculation on the synchrony of rap music and computer culture.) Indeed, the integration of phone calls into talk radio has cooled down that medium as well, and has begun to break down the non-interactive bastion of what we can now see as the old-fashioned voyeuristic coolness of TV.

At the same time, we can see McLuhan's thermostatic or reversal principle coming into play, as the pervasive coolness of electronic texts creates a need for some scraps of detail, some warmth of image, lest we shiver unduly online. The icons of open Windows and their offspring on online services are too ambiguous, cartoonish, to provide much heat in this regard. But the picturing of Web pages with literal photographs, the quest to know just what the authors of pages look like, to have explicit visual renditions of places and events described, is a dash for some shelter, a clutch for some fig leaves, an assertion of some familiar, reliable sources of comfortable hearth and detail, as the cool winds of change blow ever stronger around us.

Consideration of text – cool or hot – inevitably invokes a focus on publishing, for a text unpublished, unavailable in any way to anyone other than its author, is in a real sense not a text at all. The traditional mechanisms of publishing, hot accomplices in their extreme centralization and preference for definitive works wherever possible, were already under attack in the 1960s, when most text other than poetry was still on the hot side, and McLuhan observed that the xerox was turning every author into a publisher.

In the next chapter, we trace this development through the present time, when an author for the insignificant cost of a local phone call can publish a work to an instant, world-wide audience.

10

THE RUSTED GATEKEEPER

Thomas Gray captured the cruel discontinuity of literary merit and public recognition in his "Elegy Written in a Country Churchyard" (1751): "Full many a gem of purest ray serene, the dark unfathomed caves of ocean bear; full many a flower is born to blush unseen, and waste its sweetness on the desert air."

Some two and a quarter centuries later, on the eve of the online revolution, Joseph Agassi (1982, p. 239) wrote to his potential readers about "the inability I suffer to communicate with you prior to the consent of the editor to publish my piece, and the consent of his publisher to publish" – pinpointing the obstacle faced by any writer, prior to the Web, who sought a readership larger than friends, acquaintances, and family, or what could be reasonably reached by hand-produced copies. For to go beyond that in-person circle required the collaboration of another party in the communication process – a person or organization to reproduce and distribute the writer's work in sufficient quantity – and this meant that the writer's work in some way had to meet the approval of that crucial secondary party. The alternative was to risk ending up in Gray's graveyard pantheon, "some mute inglorious Milton."

Perhaps I found Agassi's *cri de coeur* especially moving because I was the editor whose approval he needed to publish that piece. But I was also a writer, and knew full well from that vantage point exactly the frustration he was describing. And I also knew that far more could be jeopardized in the necessity to seek the consent of publishers than just the writer's blood pressure: I knew that Karl Popper – whose *Festschrift* Agassi was contributing to and I was editing in 1982 – had nearly seen his *Open Society and Its Enemies* (1945), judged by many to be as ringing a defense of freedom as Milton's *Areopagitica*, languish unpublished because Popper's "friends in the United States…solicited an opinion from a famous authority, who decided that the book, because of its irreverence towards Aristotle…was not fit to be submitted to a publisher" (Popper, 1974, p. 95). Only the timely intervention of Ernst Gombrich and Friedrich Hayek resulted in the book's publication by Routledge (happily also the publisher of this very volume you are reading).

I also knew that McLuhan shared this jaundiced view of publishers, noting as often as he could, and with special relish, how "the Xerox makes everyone a publisher" (1977a, p. 178). He was bitterly disappointed at the speed with which Harcourt Brace Jovanovich remaindered *Take Today: The Executive as Dropout* (McLuhan & Nevitt, 1972; remaindered – allowed to go out of print – in early 1978), and the way Doubleday and he could not seem to come to terms on *Laws of Media*, which, although initially contracted for with Doubleday, was published by University of Toronto Press (and co-written with Eric McLuhan) posthumously for Marshall McLuhan in 1988.

But I also realized that photocopying would not provide the solution. For although it was inexpensive and widely accessible by the 1970s, its output looked nothing like books, newspapers, magazines, even academic journals – the media in which writers had become pleased since the advent of the printing press to present their words to the public. Thus, like so many of McLuhan's observations, the photocopier as publisher was more metaphor than reality.

But also like so many of McLuhan's observations, this metaphor was prescient.

For what I did not know as I solicited and edited manuscripts, including Agassi's, for *In Pursuit of Truth* – the Popper *Festschrift* that was published by Humanities Press in 1982 – was that some two years later, in June of 1984, to be exact, I would log on with my Kaypro II personal computer to my first online session, and therein experience the rudiment of the Web that has indeed today made everyone upon it with their own Web page a publisher.

In this chapter, we examine what that means for publishing, for writing, and the growing world of online readers.

We begin with the golden age of gatekeeping in publishing, which Agassi decried and McLuhan envisioned in ruins.

JANUS ASCENDANT

The fact that the words in a book cannot be changed – whether handwritten in a manuscript to be bound or printed for mass distribution – is ample reason that someone other than their author should examine them for error, before their cementation on paper. Prior to the printing press, the number of original in comparison to classic texts reproduced was very small, and the main agenda of the monks who made copies was to make sure that no distortions were introduced in the preservation of such precious ancient knowledge.

One might think that the printing press would have changed all of that, by rendering books far less unique. In many important respects, it did. Luther's dictum that people should read the Bible for themselves rather than seeking the mediation of the Church on matters Divine became possible via the capacity of the press to print sufficient copies, and each Bible thus read was contemplated outside of the gates of Rome. Similarly, though Galileo was induced in the court of Rome to recant his Copernican arguments that the Earth moved around the sun, his books elsewhere continued to contend otherwise. In that and many other instances, the immutability of words printed on pages – beyond even the power of the still formidable Church to amend once in circulation – dislodged gates and gatekeeping more than a millennium in standing.

But the printed regime also installed its own gatekeepers on numerous new levels, as editors and publishers were charged with underwriting everything from the factual accuracy to the spelling of the new texts that poured forth. The logic was that a mistake appearing in a hand-written copy was vastly compounded when it appeared in thousands of reproductions, and thus warranted commensurately greater attention. Thus spelling, which in the age of the handwritten manuscript was almost as variable as pronunciation, became with the printed book a nearly unbending badge of erudition.

This replacement and augmentation rather than removal of gatekeeping with the printing press was a decisive moment in media history. Although the Church per se as arch arbiter of quality in publication was dethroned, each center of printing in effect became its own Church vis-à-vis the authors it consented to publish. Initially, the prime beneficiaries of this decentralization were national states, whose monarchs set up strict controls over their fledgling presses. That model of publishing continued well into the twentieth century, with the Politburo of the Soviet Union exercising the same kind of Platonic philosopher-as-king gatekeeping as the Church hierarchy. Indeed, in the case of broadcasting and newer media, it continues in some form into the twenty-first century, as we will see in more detail below (see also Chapter 7; and see Levinson, 1997b, for details on the emergence of the press in Europe and its consequences).

In the most democratic societies – namely, America and England – the press broke free of central government, and established itself economically via advertising. In America, the First Amendment to the Constitution specifically forbade gatekeeping by government – "Congress shall make no law…abridging the freedom of…press." And although this has often been honored in the breach, especially in the twentieth century (see Levinson, 1997b, and sources cited therein), gatekeeping continued for a far more fundamental reason: the media, when not controlled by government, functioned as powerful gatekeepers in their own rights. Whereas once the Church had attempted to gatekeep all of Christendom, and then a dozen or more national states attempted to follow suit, the subsequent guaranteeing of media freedom in Bills of Rights such as the American turned out to be a safeguarding of the right not only to publish but to withhold information. Janus had not been eliminated or even diminished. He had been cloned, massively.

To some extent, the very advertising that allowed the press to be free of government control made the press's own gatekeeping inevitable. With so much advertising revenue riding on the publication of each issue of a newspaper, whose pages were carefully allotted to convey *x* amounts of news and *y* amounts of advertising, newspapers could not publish all possible news, and had to take great care in deciding what news to publish. The economy of books was different, but resulted in a similar constraint: given the cost of producing a book, ranging from paper to binding to transport, publishers had to carefully evaluate what should and should not be published.

From the point of view of the reader, newspaper gatekeeping has been the more insidious, because newspapers usually present themselves to the public as printing all news, not just the news that they have allowed to pass through their gates. Thus, *The New York Times* represents itself on its masthead as publishing "All the News That's Fit to Print"; the motto could be more truthfully rendered as "All the News That We See Fit to Print."

"Letters to the Editor" are supposed to serve as a remedy for the oversights and errors of gatekeeping: if the newspaper fails to print something important, or makes a mistake in its reporting, an alert reader can send a letter that calls attention to the problem. Unfortunately, the same space limitations that engender gatekeeping in the first place work to keep the overwhelming majority of such letters unprinted.

On November 2, 1997, for example, *The New York Times Sunday Book Review* published Mark Edmundson's review of W. Terrence Gordon's *Marshall McLuhan: Escape Into Understanding: A Biography* (1997). Though generally favorable (and justifiably so), the review contained serious errors of fact regarding McLuhan's reputation, observing that "it's faded to near extinction." The absurdity of that characterization was noted by many people (see the online discussion on the Media Ecology listserv on the Internet the following week, e.g., Wachtel, 1997), including Paul Kelly, who had just a few months earlier completed a Masters thesis, under my tutelage, that examined and extended McLuhan's work (Kelly, 1997b; see also Kelly, 1997a). Kelly sent a succinct letter to *The New York Times*, pointing out some of the many scholars and recent books that falsified Edmundson's assertion. A trip to any well-stocked bookstore would have revealed the same – works by Brand (1987), Postman (e.g., 1992), Meyrowitz (1985), and others that acknowledge McLuhan as a major influence and build upon his ideas, not to mention the dozens of scholars whose articles and texts about McLuhan are published in more academic media. (See, for example, the publications listed in the participant biographies for the "Symposium on Marshall McLuhan" held at Fordham University in New York City, March 27–8, 1998; see also the advertisement for Edwin Schlossberg's book, *Interactive Excellence*, in *The New York Times*, July 23, 1998, which says its author "takes his place as a Marshall McLuhan for the Internet Age" – hardly an example of a reputation "faded to near extinction.")

But Kelly's letter (Kelly, 1997c) was never published by *The New York Times*. Thus, readers of Edmundson's review unfamiliar with the field – the generally educated, literate reader who is presumably the *Times'* intended audience – were left with a flagrantly erroneous observation as the *Times'* last word about McLuhan in this instance. "All the News That's Fit to Print" alas did not include the mountain of contradicting evidence.

In all fairness to *The New York Times*, book reviews are usually controversial regarding their opinions – such as, McLuhan is an infuriating writer – not facts, although targets of unfavorable opinions are prone to regard them as errors of fact when offering objection. Perhaps the editor of the *New York Times Sunday Book Review* who received Kelly's letter mistook it for a disputation of opinion rather than a highlighting of factual error. But the deeper problem remains: with a dozen or more books accorded lengthy review each Sunday in the *Times*, there is no way that it or indeed any newspaper or magazine could possibly publish letters correcting errors of fact in every or even most of the reviews. Publication of such gatekeeping of gatekeeping would require a separate newspaper or magazine of its own. One hopes, of course, that reviews never contain errors of fact. But given that they, and all work done by human beings, inevitably do on frequent occasion, the best we can expect in "Letters to the Editor" is but a trickle of correction – and the often mistaken implication that reviews uncorrected by such "Letters" are completely accurate.

Thus, the economy and physical constraints of print on paper conspire to keep the gatekeeper ascendant. And classic network broadcasting is even worse in this regard, being more expensive to produce than a newspaper or magazine, and locked into an ironclad time restraint. Pages can be added to a newspaper or magazine, especially if there is accompanying advertising, but only the gravest news event warrants extension of a newscast beyond its allotted thirty or sixty minutes. And yet Walter Cronkite concluded each of his CBS-TV nightly newscasts in the 1960s and 70s with a sonorous "And that's the way it was" – the broadcast equivalent of "All the News That's Fit to Print," and just as misleading. A more accurate tagline would have been, "And that's the way the editors at CBS decided you should think it was."

Today, television networks present their news with much the same underlying attitude as Walter Cronkite, if without the explicit (yet historically endearing) self-benediction. But call-in programs on radio and cable television, and the rapid expansion of cable television in general, have begun to pry open the gates. These, far more than the Xerox cited by McLuhan, are actually allowing samples of "everyone" to publish on the air.

And then there is the Internet, which in making everyone with a Web page a publisher – in creating an environment where costs of paper, binding, transport, and broadcasting all vanish – stands poised to tear down the gates entirely.

One might expect that authors heretofore kept behind the gate, or obliged to pay its toll in seeking approval of an army of keepers including agents, editors, book reviewers, and bookstore managers, would rise in joy at the Internet prospect.

Instead many, including me (after all, you are reading this in printed book form), apparently have come to crave the ministrations of our gatekeepers, much as some prisoners come to love their jailors.

This persistence of the assumption that gatekeeping is needed may be its most enduring legacy, and one that survives the advent of media like the Internet that make it unnecessary.

GATEKEEPING MENTALITY

The logic of gatekeeping – whether by Church, national states, or the media themselves – is that information is like a food or a drug, which, apropos the Pure Food and Drug Act of the United States and similar laws in most countries, requires inspection or certification before it can be made available to the public. To offer information unvetted is, on this reasoning, to risk poisoning the public, as it could be from spoiled food or bogus medication. Thus, Justice Oliver Wendell Holmes, Jr., writing the unanimous opinion for the Supreme Court in *Schenck v. United States* in 1919, contended that "the most stringent protection of free speech would not protect a man from falsely shouting fire in a crowded theater, and causing a panic...the question in every case is whether the words...create a clear and present danger..." (see Tedford, 1985, pp. 70–1).

Ironically, Holmes intended such a seemingly reasonable criterion to be the sole gatekeeping function of government in communication, his being otherwise a staunch defender of the First Amendment. He had little time to wait to see the error of his reasoning. In November, 1919 – eight months after the *Schenck* ruling in March of that year – Holmes found himself in the minority (along with the also well-intentioned Brandeis) in *Abrams v. United States*, in which the Supreme Court ruled that a leaflet Holmes regarded as "silly" posed just such a clear and present danger (see Tedford, 1985, p. 75). Thus, Holmes' attempt to open the gate to gatekeeping just a little, then limit it, had just the reverse effect: with the First Amendment as sentinel now severely compromised, the gate to gatekeeping of media in the United States was now wide open. Although subsequent Supreme Court decisions in the twentieth century would in general support the right of newspapers and the press to publish without government interference (the Pentagon Papers case, in which the Supreme Court unanimously rejected President Richard Nixon's attempt to restrain *The New York Times* and *The Washington Post*, was the most vivid example), broadcast media have fared much more poorly. Thus, the Federal Communications Commission has acted as a quite exacting gatekeeper, dispensing broadcast licenses to radio and television stations – tickets to pass through the gate – on the basis of whether they broadcast "in the public interest," since the Federal Communications Act of 1934 (see Levinson, 1997b, and sources cited therein for details). As we saw in Chapter 7, the Communications Decency Act of 1996 attempted even greater gatekeeping for the Internet – providing for up to two years in prison and a hefty fine for violators of its provisions against "indecent" and "offensive" material online – but the Supreme Court, in effect judging the Internet to be more a newspaper than a TV station, struck down the law as unconstitutional in 1997. Almost needless to say, the mentality of gatekeeping persists in Congress.

Works of science and scholarship and literature, presumably less immediately toxic to the public weal than evil political tracts and broadcast entertainment, have by and large not been regulated by any democratic government. For these modes of communication, the media are supposed to gatekeep themselves. Thus, in the case of scholarly publications, respected journals distribute potential articles to "outside" readers (i.e., readers not connected to the journal), who are asked to render a judgement as to their value; this is usually done "double-blind," i.e., the authors do not know the names of the readers, and vice versa. Similar procedures are often followed by publishers of scholarly books (Routledge sent my initial proposals, along with sample chapters, for both the present book and *The Soft Edge* to outside readers).

Since so little if any scholarly material is published without such gatekeeping, judging its efficacy is difficult. Scams such as the Piltdown Man, or Sir Cyril Burt's studies of twins with fabricated data, however, show that experts at the gates offer no guarantees: these frauds took decades to uncover, despite the best attentions of presumably the best minds in those fields of study. How many similar deceptions have escaped the ministrations of gatekeepers?

The arts – music, literature, painting, cinema, etc. – entail different principals and constituencies, yet admit to the same perils of gatekeeping. The American government's attempt to improve the quality of music played on the radio in the late 1950s via its investigation of "payola" – or the playing of records because disc jockeys were bribed to do so, rather than because the DJs thought the music was good – resulted in the dullest period of rock 'n' roll history, 1960–1963. The Beatles came to the rescue at the end of that year, having themselves been denied contracts by more than twenty recording companies – another debacle of gatekeeping –

before EMI Records had the good sense and ears to sign them. Cinema was similarly hobbled by the notorious self-imposed "Code" in America that prohibited such expletives as "hell" and portrayal of even married couples sleeping in double beds (the British code, also self-imposed by the British Board of Censors rather than the government, was even stricter).

In all of these cases of gatekeeping – whether by government, scholarly journal, record company, film board – the question which McLuhan's "everyone a publisher" invites us to consider by implication is: What, on balance, is actually gained? Does the ineffective keeping of trash and error from the public – certainly not poison, let us admit, but let us also accept, for the sake of argument, that it has no redeeming political, scholarly, or artistic merit – does its filtration out of our communication systems offset in benefit the damage done when the same filtration deprives us of the Beatles, a Karl Popper, an unsung Milton, a gem of purest ray serene?

I would argue that the risk of losing any of these – and we came perilously close with Karl Popper and the Beatles – far outweighs the unappraised "glut of possibilities" which Michael Heim (1987, p. 212), early in the online age, saw as one of its main potential drawbacks.

But it turns out that our choice in the digital age may be better, in that we may be able to do away with gatekeeping and maintain the appraisal, or transform the gate from a filter to a valuator.

FILTRATION VERSUS EVALUATION

Although the gatekeeping of Church, government, and private media are all intended to keep noxious information from us – or ensure that we only receive information deemed beneficial – there is a significant difference in the ultimate motivation of gatekeeping self-imposed by the media. Whereas Church and state begin with an ideology – a conception, however unclear, of what should be kept from the public (e.g., obscenity) – the media begin with the technological and economic reality that they can only publish a small amount of all potential information, and then seek to justify this constraint ex post facto with an ideology (e.g., we only publish information that is newsworthy).

If this analysis is correct, we would expect government to view the liberation of text from paper as just another occasion for exercise of its ideologically driven (protect the public from danger) gatekeeping; the advent of new media, after all, in no way contradicts the ideology, and indeed gives it even more urgency. Thus, the Communications Decency

Act of 1996 was the most regressive in terms of scope and punishment since the Sedition Act of 1798.

But if the above analysis is correct, we should also expect the media to be fundamentally altered in their gatekeeping by the vast publication possibilities of the Web – for these possibilities break the technological and economic bottlenecks of print on paper (and broadcasting on the airwaves), and thus knock the props out from under the media's rationale for gatekeeping.

The success of Amazon.com as an online bookstore – by 1998 the third biggest bookseller in the world (Nee, 1998) – may be the first evidence of such a result. Unlike in-person bookstores, which due to their own physical constraints cannot possibly have every book in print on their shelves, Amazon.com lists for sale just about any book available from any offline publisher. True, a potential buyer could walk into any bookstore and order any book in print, but such shipments often take weeks to arrive, and, besides, the display of some but not all books in print in a bookstore is a kind of gatekeeping that obviously favors the books displayed on shelves. In contrast, books are more equal in their virtual display and availability on Amazon.com (not entirely equal, because some books are displayed more vividly than others, and can be shipped with greater speed – but certainly more equal than books displayed or not displayed at all in in-person bookstores).

Amazon.com's approach to book reviews is also indicative of a new gatekeeping without gates. Their editors select a few books for special recommendation and review, and they also publish online, alongside of the book listing, reviews selected from major media such as *The New York Times*. But Amazon.com also solicits and publishes reviews from all readers, as well as commentary on books by their own authors. Thus, although I suffered to read on Amazon.com a *Kirkus* review of *The Soft Edge* (1997b) that was factually inaccurate, I was also able to provide my own – one hopes, more accurate – summary of the book. The Web bookstore thereby provides guidance and evaluation without shutting the door on opposing information. In contrast, BarnesandNoble.com (the online bookstore of Barnes and Noble), unable to completely shed its offline restrictive gatekeeping, prides itself on publishing only "professional" reviews in its book listing. And, as we saw above, completely offline venues such as *The New York Times Book Review* have nearly no facility for publishing contrary views (just a minuscule and highly selective "Letters to the Editor" column).

Indeed, the same freedom from paper constraints that makes

Amazon.com a non-gatekeeping gatekeeper should have equivalent effects on the publishing process as a whole. Magazines, journals, and books will and should continue to be published, on and offline, as expressions of the publisher's taste and editorial acumen; these are the equivalent of Amazon.com's "Editor's Choices." But the capacity of anyone with a Web page to therein publish his or her own story, essay, book means that texts made available with a publisher's imprimatur will no longer lock such non-endorsed texts out of the public arena. The Web allows gatekeeping both on it and in society as a whole without filtration. Instead, gatekeeping becomes endorsement without punishment.

Of course, a book published with the endorsement of attractive hard covers will likely have an edge, certainly in traditional bookstore sales, over an equivalent text published only online. (And note the policy of *The New York Times* to give major reviews only to hardcovers, not even paperback books – due, again, to unavoidable space constraints.) But as more texts of all sorts become available in primary form online, where an author working from a personal computer at home could conceivably format and frame the online text to look as appealing as online text packaged by a professional publisher, the aesthetic advantage of professionally presented texts may well diminish.

And there is another significant advantage to online endorsements: they can be "intelligent" – pro-active or "matchmaking" – to potential readers.

FROM GATEKEEPER TO MATCHMAKER

Amazon.com seems further instructive of how a digital future might operate along the lines of McLuhan's "everyone a publisher." Listings for books frequently show titles of similar books that buyers of the instanced book also purchased. Buyers are encouraged to "click" on options that cause Amazon.com to notify them when future books by a specified author or about a specified topic become available. These and similar "push" technologies – i.e., programs that, once set, automatically bring selections to customers, in contrast to "pull" technologies which require the user to search the online bookstore or Web anew for each desired selection – show the transformation of gatekeeping in online bookstores to special delivering. Not only do online "gates" no longer keep texts from potential readers, they have metamorphosed into winged butterflies that alert readers to the availability of books that might otherwise have been destined to blush unseen.

Such "matchmaking" (as Stanley Schmidt, editor of *Analog: Science Fiction and Fact* magazine, aptly calls this new editorial function; see Levinson & Schmidt, forthcoming; also Schmidt, 1989) can further develop in a variety of ways. Alexandria Digital Literature, for example, invites its online customers to participate in a survey that seeks to gauge their tastes in fiction; once such profiles are established, they are used to guide readers to stories in the online library that are likely to be of interest. Of course, readers can select any stories they please – they are free to ignore the choices of the library program – and thus these "gates" are entirely unrestrictive.

Nor need the mechanisms of such matchmaking be limited to silicon circuitry. Although Alexandria Digital Literature operates on artificial intelligence – a sophisticated computer program (see its Web site at http://www.alexlit.com) – good old-fashioned human intelligence can certainly do the same thing, in the person of an editor who makes selections for his or her readers. Indeed, that is what editors have in effect been doing for print publications all along. As Schmidt (1989, p. 5) observes to his readers, "Very few of you would be willing or able to get through the 500 or so science fiction stories I read in a month to find half a dozen I think you'll like....So you hire me to find good stories for you." Again, the difference between Schmidt or any editor performing a similar service in an online environment in which any author can publish for the price of a phone call to a computer node connected to a Web page is that such online editorial selection or privileging of a small group of stories, while disprivileging the huge remainder of stories by not bringing them to the attention of specified readers, does not consign them to virtual non-existence in the public realm by leaving them unpublished. The online editor thus becomes an endorser rather than a door dragon, as the pernicious process of filtration is severed from the classic editorial mandate.

Nonetheless, in a society whose publishing economy is derived and still strongly rooted in paper and its relative scarcity, shifts to a digital environment of informational plenty will not be easy. In an online world in which "everyone is a publisher," who will pay the editors for their matchmaking? In the traditional paper world, publishers paid editors for their labors. Will "everyone" – authors, readers – be in a position to pay editors online? Indeed, as Schmidt (1989) points out, the same technology and logic that makes everyone a publisher also can make everyone an editor online. And, in fact, the Web is now brimming with pages that list books and stories and restaurants and many things favored and recommended by the page host.

Consideration of online economies brings us into contact with aspects of McLuhan's ideas that go beyond publishing per se. We have already encountered some of this in the ways his notions of the global village (Chapter 6) and centers everywhere and nowhere (Chapter 7) are becoming real in the digital age, like bodies glistening into full flesh-and-blood existence as they emerge from being "beamed" on *Star Trek*. In the next chapter, we turn to another facet of labor in the new millennium, especially pertinent to writing, editing, and publishing because they seem like fish returning to water online. We look at how, decades before the personal computer and its integration of so many activities on its screen, McLuhan investigated the electronic proximity of work to…play.

11

SERFS TO SURF

As Joshua Meyrowitz demonstrates in *No Sense of Place* (1985), a key element in McLuhan's understanding of electronic media is the way that they blur social distinctions and categories engendered or at least supported by print. McLuhan's notion of electronic media exploding centers (see Chapter 7 in the present book) is one expression of this, and Meyrowitz explores the electronic erosion – with television as the main dissolving agent – of such seemingly fundamental distinctions as public versus private, and adult versus child. The active ingredient in this electronic solvent is access: even a pre-literate young child can see the same news on television as her parent, and that news, especially when it is about the President of the United States, can be far more personal and private than what used to be reported in newspapers prior to the advent of TV.

McLuhan early on caught another aspect of this electronic blurring of boundaries when he noted that "Heidegger surf-boards along the electronic waves" (1962, p. 295). McLuhan was comparing the "standing reserve" of technology in Heidegger's philosophy to the waves that the surfer stands upon and is moved by – the surfer's moving ground – and relating this all to the non-locality of electronic media. McLuhan also connected this standing motion to "the physicists" (e.g., M. McLuhan & E. McLuhan, 1988, p. 63), as in the restless placelessness of Heisenberg's uncertainty principle – the position and momentum of a subatomic particle cannot both be precisely known at the same time – and their investigation of "standing waves." But what jumps out about the invocation of "surfing" from the vantage point of the new millennium is its widespread adoption as a metaphor for browsing the Web.

Surfing the waves of the ocean, as anyone who has tried to do it even casually can attest, is a serious business if you want to do it right – if you want to stand for more than a moment without crashing. It takes concentration and practice, talent and savvy – much like riding a motorcycle, or even a fast car with the convertible-top down. "Poe's mariner in 'The Descent into the Maelstrom,' " McLuhan observed, "staved off disaster by understanding the action of the whirlpool" (McLuhan & Fiore, 1967, n.p.). But like all of these activities, surfing is also fun. Indeed, fun – "the amusement born of rational detachment," as McLuhan observes of Poe's mariner – may be key to successful negotiation of this whirlpool.

The same duality of business and pleasure – of particle particularity and wave emotion – pertains to surfing the Web. The notion of doing business upon it is sometimes an oxymoron because it can be so much fun just being there. Yet crashes and commercial devastation await those who get too far ahead of the wave. *WIRED*'s online alter-ego, *HotWired*, lost so much money after *WIRED*'s initial success that the whole magazine operation was sold to Condé Nast in 1998, absent the magazine's founder and guiding light, Louis Rossato (see Wolff, 1998).

Here we look at surfing on the Web, and the likely ways it will continue to combine work and play in the new millennium. We will see that, in transforming both activities in the combination, the online milieu holds new opportunities for human expression and satisfaction in the business and personal realms, as well as pitfalls for business conducted more personally, and families exposed to business via personal computers in the home twenty-four hours a day.

We begin, as usual, with McLuhan's perception of this transformation in media long before computers.

INSIDE-OUT ON THE PHONE

In an age just prior to cell-phones, McLuhan observed that "the North American goes outside to be alone...the North American car is designed and used for privacy" (1976, pp. 48, 50). The reason for this, of course, was that electronic media – the telephone especially – had already long evicted people from the privacy of their homes. Today, barrages of telephone calls offering special insurance deals, new calling plans, and all manner of business at all hours of the day and evening and on weekends bring home McLuhan's point as never before. And that is before we take into account the public faces of television and computer screens in just about every room in the home. Even equipped with a car phone, the automobile on the road is still the most private place to be, informationally, for the only people who can reach us there and require our response are those we have entrusted with our car-phone number.

As we enter the twenty-first century, likely few among us have any conception of what the home was like prior to the phone. Although the telegraph sped its messages at the speed of light, it strictly kept its distance with a knock on the door for delivery and a trip to the telegraph office to commence the transmission. Books, newspapers, and periodicals were of course brought into the home, but the people who took them within their shelter knew beforehand more or less what they were conveying, and once inside, there was no chance that the book or newspaper would ring for attention.

The telephone, first called the "talking telegraph," demanded our attention not only with its sweet siren ring, but because that ring promised to have a live person at the other end. Moreover, the person's identity at the time of the ring was unknown – for the first century or more of the telephone, prior to caller ID – which made the ring all the more compelling. After all, our caller could be wanting to clinch a crucial business deal, or profess an everlasting love for us, even if most of the time the call turned out to be an appeal for a charitable contribution or an invitation to dinner at the in-laws. Radio and television would bring other important pieces of the public into our private homes – notably, the more organized aspects of public life, such as politics, and the formal audio and audio-visual commercials of business – but unlike the telephone, they did not ring when they were off. Turning on the radio or TV was the equivalent of making a decision to bring a newspaper or book into the home, although the audience for the electronic medium usually had far less explicit knowledge of its content beforehand than did the reader.

But the telephone's unique dangling of personal hope and real interaction with the outside world would have made its ring the most irresistible even had radio and television turned themselves on unexpectedly throughout the day, of their own accord. (And, on that regard, McLuhan may have paid too much attention to television or not enough to the phone in their respective deprivatizations of the home.) And the telephone had something else, even more seductive in many ways than the ring, that helped turn the home inside out, or into a place of business: we could initiate business *from* the home by the simple expedient of making a call.

Of course, people took work home from their shops and places of business long before the phone. Indeed, farmers worked at home for millennia, and still do. But the work in those and related cases entailed a human being working with a product – the papers, the ledgers, the crops, the livestock – and not vis-à-vis other people. In that sense, that kind of home work is akin to homework from school, preparatory but not consummative to the endeavor. For the farmer to consummate the farmwork, crops must be hauled to market, just as the student (prior to online education) had to attend class to be tested and to receive a grade. Groups of people may work together on a farm, but the consummation of the enterprise is still off-farm – or, in today's parlance, offline.

The initiation of business from the home via a telephone call is revolutionary because there is a live person not a product at the other end of the line, and thus business can be conducted in its entirety on the phone. Sales pitches can be made that result in sales (whether by checks in the mail, or later by credit card), deals can be consummated, all on the phone. Thus, as the Industrial Revolution went into high gear, drawing farmers into factories, and others into office buildings to work as secretaries and chief executives and a myriad of jobs in between but equally away from the home, the telephone – quietly, as an initiator of calls, more than loudly, as the receiver of calls – christened the literal home office as a mass phenomenon.

In the final quarter of the twentieth century, the fax and the modemized personal computer – both, not coincidentally, utilizing the same lines for telecommunication as the telephone – transformed the purely acoustic office in the home into a more full-bodied reality that includes paper, text, pictures, and moving images.

THE HOME OFFICE: FRIEND OF THE FAMILY

We have already discussed Connected Education, Inc. – the educational corporation my wife and I founded in 1985, to offer graduate courses completely online, in cooperation with major universities – and its relevance to education in a world of diminishing centers (Chapter 7). Since nearly one-hundred percent of its business was conducted online from first Kaypro CP/M and then DOS personal computers in our dens connected via modem to our phone lines – the only necessary business outside of the office being the occasional meeting with the appropriate dean or departmental chair I was obliged to attend – "Connect Ed" (our trademarked nickname for the organization) also provides a textbook study of the home office in the early days of the digital age. Here is how it happened...

Our Fall 1985 inaugural term attracted a grand total of twelve students. But one took his entire course from his home in Singapore, another from his office in Tokyo, and several from places around the United States other than New York (headquarters of both Connected Education, and the New School for Social Research, our university partner at that time). We were thus convinced that online education was economically viable; by the Spring of 1986 our enrollments were exceeding fifty. I soon after resigned my tenured associate-professorship at Fairleigh Dickinson University in Teaneck, New Jersey (about ten minutes across the Hudson River from New York), and staked the financial well-being of my growing family on Connected Education.

Only three years earlier, not only had I been gainfully employed in a tenured job that took me out of the home, but my wife Tina had been Promotions Manager at Smith-Sternau, an insurance administration company with offices in the heart of Manhattan. In the Fall of 1983, the most sophisticated telecommunications technology that either of us had used was the telephone; our "word processors" were IBM – that is, IBM Selectric 2 typewriters. We had no children. All of that was about to change.

Simon Vozick-Levinson was born in November 1983. Tina took an extended maternity leave, fully intending to return. But the digital revolution was soon to offer other opportunities for work – which would not conflict with the time she wanted to spend with our son. In June 1984 I sat down with my first computer and modem, and logged onto the Electronic Information Exchange System ("EIES," operating on a central computer at the New Jersey Institute of Technology), in preparation for an online course I would be teaching for the Western Behavioral Sciences Institute in September (see Chapter 9 for details). The first night I logged on – in June – was enough to teach me the value of online communication for scholarly discourse: I sent an e-mail to Langdon Winner (author of *Autonomous Technology*, 1977), whose name I noticed online, and received a reply in an hour (Langdon was in California, and I in New York). The first day of the online course I taught – in September – was enough to teach me the value of online education. By the end of that year, I had submitted a proposal to the New School for Social Research ("The New School Online," 1984) for an online academic program. The New School accepted the proposal in January 1985, and Connected Education, Inc. was off (or online) and running.

The Spring, Summer, and Fall terms of 1986 made for a halcyon year for Connected Education, and in November 1986 our family grew again too with the birth of Molly Vozick-Levinson. The scene in the Vozick-Levinson household, also Connect Ed central, was a home-office ideal: Working from home, we were able to give our 3-year old son and newborn daughter the attention they needed and deserved. And the asynchronous nature of our work online – we logged on at times of our choosing, and could log off any time we wanted as well – meant that although our work for Connected Education could and certainly was interrupted many times by the children, no one at the other end of the computer connection could know that. Tina would just log on at another time, and resume the e-mail she was writing, and I would do the same for an online course I was teaching.

Moreover, since the Kaypros we were using at first as our personal computers were "transportable" (weighing under 25 pounds – no problem at all to lug around for anyone used to carrying toddlers), we were able to move our home office every summer up to the cottage we rented by the bay on Cape Cod. Many was the July evening I would come back from the water, dripping wet, and log on in my bathing suit to enter a comment in an online course with a thought that had occurred to me as I was floating in the bay near sunset. Although the waves on the bay were not high enough to surf, I was a serf going surfing in every other way those days. McLuhan, I'm certain, would have been pleased. (Marshall and Eric had been at our home for dinner in 1978 – Tina afterwards named that pot roast recipe the "McLuhan pot roast" – and Tina and I had wonderful dinners with Corinne and Marshall at their home in Wychwood Park several times. We always keenly regretted – and still do – that Marshall did not live to meet our children.)

Thus, the digital home office thoroughly blurred the boundaries between work and play in our case. In an environment in which business and family were so fully integrated, in which we – as parents and proprietors of the business – could shift from one role to another quite easily and often instantly, we found ourselves with much greater control over our lives. Freed from the tyranny of having to work on pre-set schedules – which, moreover, took us out of the home and away from people we loved – we found the digital home office a profound liberation.

Again, McLuhan likely would have agreed, writing of "the kind of slavery involved in a repetitive operation like publishing a journal" (McLuhan in Stearn, 1967, p. 270). No friend of the tyranny of schedules for anyone, he added that "most readers of most journals are very unhappy about their regular appearance."

But has the liberation of the serf online been unqualified?

THE HOME OFFICE AS FRIEND: COUNTER-INDICATIONS

In considering the advisability of transforming any social structure, we need to take into account not only its disadvantages.

What benefits are conveyed in the boundaries between work and play that the home office erases?

The main advantage of any pre-set schedule, or boundary beyond our ability to easily manipulate, is that it allows us to reap the rewards of such divisions without having to create the divisions anew each time. In a nine-to-five job that takes us out of the home, we know – without having to accord the matter any additional thought or attention – that when we are in the job environment we are there to work. Certainly we can take a personal phone call, but only the most extraordinary news could pull us completely out of our work. Similarly, although a business call at home could intrude upon our family and personal time, we generally know with assurance when working in an office that when we are at home we can give uncontested attention to our family and personal affairs.

The key point here is that it is the enforced separation – the working in an office *away* from home – that gives us this informational security *in* the home, and vice versa. The home office allows such security, but puts it entirely in the hands of our moment-to-moment choice, rather than the assured pre-set schedule. And, life being what it is, sometimes the moment-to-moment choices are difficult to implement.

As Simon and Molly grew older, many is the time I would bellow from my den while on the computer, "Could you keep it down? I'm trying to work!" Holding a 6-month-old on your shoulder or a 2-year-old on your lap is a lot easier than driving a 10-year-old to a birthday party when you're surfing the Web.

The result of such lumpy integrations of family and business is that both can suffer – family time can be corroded by work, which in turn can be compromised by family interruptions.

What, then, determines whether the home office's advantages outweigh these hazards?

The answer likely resides in the natures of the family and the business. Younger children have the happy coincidence, from the point of view of the home office, of both requiring the greatest amount of direct parental involvement yet being the easiest to integrate into a work-at-home schedule. The work of being a writer – at least for me – requires my having access to a computer at all waking hours, if possible, because I never know when inspiration will hit me, or when a planned hour-long session will turn out to be a bonanza day in which I write 8,000 words. With that kind of "job" (and I put that in quotes, because writing when it goes well can be incredible fun), any other work that takes me out of the home, away from the computer, can be destructive. Thus, we might say that a writer with young children is an ideal candidate for working at home.

Of course, apropos McLuhan's prediction of "centers everywhere" (see Chapter 7), personal computers are located not only in homes but in offices, where they provide similar opportunities for flexible work. I teach at Fordham University – which takes me out of the home several days a week – but the personal computer in my office allows me to continue my work as a writer without missing a beat. In that sense, the personal computer is more integrating and blurring of distinctions than earlier electronic media such as television, for the computer transforms the social place in which it happens to reside – home or office – into a milieu for pursuit of whatever specific work is done on the computer. The personal computer, in other words, sets the work agenda, or establishes the nature of the business transacted in the place. This is the case not only for word processing, but, obviously, for connecting to the Internet. Connect Ed online students often participated in their courses during lunch breaks from their places of business.

But if the personal computer, as a vehicle of work, blurs boundaries between office and home, between office and classroom, between office and den, between office and office, it also seems to have an intrinsic quality which often lends a touch of play to the tasks we accomplish upon it. The commercial success of the Macintosh for more than a decade attests to this whimsical aspect, as does the triumph of Windows. It is almost as if there is something in the personal computer per se that appeals to our sense of play.

As we will see in the next section, it may be a quality which, while embodied in specific ways in the computer, is attendant to all media when they are first introduced into society.

TOY, MIRROR, AND ART

Apropos mixtures of business and pleasure, my first published major scholarly article, "Toy, Mirror, and Art: The Metamorphosis of Technological Culture" (Levinson, 1977b), was essentially written on my honeymoon in 1976, on a portable Smith-Corona typewriter in our hotel room, as part of my "Candidacy Exam" in the Media Ecology Ph.D. program at New York University (where I earned my Ph.D. three years later). It turned out to be my most popular scholarly article, having been reprinted four times in the ensuing years (see Bibliographic listing under Levinson, 1977b, for details). Only my science fiction – indeed, just one story at that – has enjoyed anything close to such reprint success (see Bibliographic listing under Levinson, 1995a, for details).

Like McLuhan's corpus, "Toy, Mirror, and Art" is mostly innocent of the personal computer revolution. But it does contain a significant reference to a piece in the *New York Post* (Keepnews, 1976) which "reports that sales of home computers are spreading 'like wildfire,' and mostly to 'techno-fetishists' who play a variety of visual and intellectual games with computers" (Levinson, 1977b, p. 164).

Indeed, one of the theses of "Toy, Mirror, and Art" is that media tend to make their grand entrances into society at large as toys – as a gadget or gimmick that people appreciate just for the fun of it, not for the work that it may accomplish. William Orton, President of the Western Union Telegraph Company, advised his hapless friend Chauncey Depew in 1881 not to purchase one-sixth of Bell Telephone stock into perpetuity for $10,000 because, in Orton's view, the "invention was a toy" with no "commercial possibilities" (Hogarth, 1926); and John Brooks (1976, p. 92) relates that use of the telephone in England was delayed for at least a decade due to the conviction that it was just a "scientific toy." What Orton et al. missed, of course, is that the very playfulness accorded the phone would act as a lubricant for its widespread acceptance and use – we become comfortable with things with which we play – and render it a phenomenal commercial success. By the turn of the nineteenth century, the number of phone calls made in America would exceed the number of telegrams by a factor of better than 50 to 1 (Gibson, n.d., p. 73).

Similarly festive receptions awaited other media. Edison's phonograph was initially promoted as a novelty despite the inventor's more serious expectations (Josephson, 1959, p. 172). The first motion pictures were enjoyed not for their story lines, which were nil, but for the sheer thrill of seeing people come to life on the kinetoscope or screen – hence, such pioneering hits as *Fred Ott's Sneeze* (Edison) and *Baby's First Meal* (the Lumières). Indeed, back in the first millennium, the Chinese invention of the printing press was used primarily for dissemination of Buddhist pictures and holiday greetings, with no outpouring of news and science as would later happen when the press was adopted and adapted in the West as an engine of mass communication.

The Chinese example is especially instructive because it shows that the transition from play to work – from taking pleasure in the general process of the toy to focusing on what specifically the device does for us, from being amused and amazed by its very existence to concentrating on its content, as when we look in a mirror – is by no means assured or similarly paced for every medium. McLuhan (1962, p. 185) attributed the Chinese failure to develop a mass press to the impracticability of setting type in an ideographic system of 5 to 20 thousand characters; William McNeill (1982, p. 49) saw the Chinese discomfort with reaping profits as stymieing the "autocatalytic" economic/technological forces that arose in Europe in the twelfth century and after. Both are likely right, and perhaps other blocking agents played a role as well, but the larger point is that, for whatever reasons, a device which within decades after its introduction helped transform Europe into the modern world had almost no discernible effect for century upon century in the home of its invention, in China.

Not that all play and no work is necessarily deleterious or undesirable for every technology. The Chinese also invented gunpowder and rocketry, and at first devoted their use to holidays, like Fourth of July fireworks in America. And although I see exploration and settlement of the universe beyond Earth as crucial to our species, and of course recognize our current reliance on rocketry to accomplish this, I also would not mind in the least had those Chinese toys never matured into the handguns, rifles, and missiles that continue to beset our lives.

Nonetheless, cases of arrested development in technologies used for the transmission of information – or even misperceptions of their potential – have usually proven not for the best. Orton's myopia regarding the telephone not only cost his friend dearly, but helped set his mighty telegraph company on a course that would eventually see it purchased and all but consumed by Bell's telephone company. Radio, the next telecommunications medium, followed a similar pattern. Although Marconi intended his wireless to provide life-and-death service on ship-to-ship transmissions, the wireless operator on the *Titanic* shut out a crucial message from the *Californian*, warning of treacherous ice conditions, because he was overwhelmed with transmission of vacation greetings from *Titanic*'s passengers. Three years after the *Titanic* sank, David Sarnoff – ironically, a wireless operator on duty in New York at the time of the tragedy – proposed that radio receivers could be located in every home, where they could serve as receivers for musical entertainment. (It was almost as if Sarnoff, having witnessed first-hand the partial failure of the wireless as a practical instrument – the *Carpathian*, reached by the wireless, of course did manage to save almost a third of the passengers – needed to assert the more comfortable application of this new instrument as a medium of amusement.) This indeed became the rage for radio in the 1920s and 30s (and still is). But it left the world unprepared for Hitler's and Goebbels' deft eventual use of the medium as a very serious vehicle of propaganda – fortunately offset by the equivalent broadcast talent of Churchill and Roosevelt (and, regarding the rallying of his country against Nazi invasion, Stalin).

The question concerning computers as objects of play and vehicles of work is not if they are both – they indisputably are, and help us express both aspects of our being – but what is the optimal mixture, and when are we likely to achieve it? Clearly, word processing and Web surfing are now more prevalent in public awareness about computers than games. Word processing, especially, is a working transaction with reality in which the content – the writing created – is paramount (again, like the face in the mirror). But Windows still comes bundled with games. And the Web is indeed often surfed for the fun of it.

Children are both a cause and a beneficiary of the continuing playfulness of computers. Although media mature into practical mirrors in the overall society, they are taken anew as toys by each new individual who receives them for the first time. Thus, from the point of view of the child, the telephone, the answering machine, and the computer are all toys at one time or another – media forever recapture their youth in the first vision of children. But the proximity of serious work and games on the same computer encourages the child who has come to know the computer as a toy to use it for word processing or conducting research for school on the Web. Unlike the automobile, in which the child must remain a passive passenger to its operation, the computer facilitates the child's passage into informational adulthood when the child is still a child. Many adults, including many in government, equate such access to information with pornography, and seek to restrict it (hence, the unconstitutional Communications Decency Act of 1996). I argue, in contrast, that the fear of pornography and its damage is groundless (see Levinson, 1997b), and that access of children to the worlds of other knowledge on the Web is the biggest boon to education since the printed book.

Which is not to say that I think stalking of children and other criminal activities online ought not to be prevented and punished to the fullest extent of the law. Indeed, I have proposed in my talks that "Megan's Law" – the posting in public places of pictures of people convicted of offenses against children – be extended to posting of those pictures online, where they can be easily accessible to parents anywhere in the world. But pursuit of criminals online, and policing of online criminality, are activities quite distinct from attempting to regulate pornography.

Consideration of pornography online brings to the fore another facet of computers and their performance as toys, mirrors – and art. On the one hand, computers are a mixed bag of toys (used for fun) and mirrors (used for practical tasks). Indeed, in as much as games, word processors, and Web browsers are separate programs, we can consider these separate media – another example of McLuhan's media within media (see Chapter 3), each of which has its own toy/mirror valence. Further, as computers continue to evolve into new programming possibilities, we can expect each new program to commence in our use with some aspects of the toy. That is why newer kinds of usage, such as Web browsing, are today more playful than word processing, which has been around a lot longer (and newer kinds of Web browsing, like Java, are more playful still).

But the public perception and use of new media also develops – or grows up – in another way. Pornography, if you think about it, is neither quite a playful toy nor a practical, workaday undertaking. It is rather something else – something post-toy and post-work – something I would call art, in this structural sense, albeit a controversial example of it.

The progression from toy to mirror to art can be seen most clearly in the history of motion pictures. The movies begin, as indicated above, with *Fred Ott's Sneeze* and *Baby's First Meal*, in which the content is insignificant and the amazement of seeing anything in motion on a medium is everything. But soon, the same Lumière Brothers who made *Baby's First Meal* show *The Train Enters the Station* in their theater in Paris. And something even more amazing happens: people in the audience duck and scream as if the train were chugging right in at *them*. In some part of their brains, Coleridge's willing suspension of disbelief has kicked in, and the audience is responding to the content of what they see on the screen as if it was real – a mirror of reality. Of course, the audience enjoys this experience – and thus the fun continues – but it is an enjoyment more of the content, as if it was really happening, and less of the technique which is bringing us the content, no longer regarded as an amazing gimmick. It is a very different, more serious, kind of fun. (The shift occurred in the development of radio as entertainment as well: as the build-it-yourself, crystal radio fad waned at the end of the 1920s, listeners began focusing more on the music and less on the magic of the instrument that conveyed the music.)

But the evolution of movies did not stop there either. Within a few years, Georges Méliès uses a Lumière Brothers camera (he is obliged to bribe a night watchman to obtain it, after the Lumières refuse to rent it to him) on a boulevard in Paris one bright afternoon. The camera jams; he resumes filming when he fixes the jam; he develops the film, projects it, and discovers the principle of editing: although the sections before and after the camera are not connected in real time (minutes passed while Méliès unjammed the camera), they flow one into another like magic, as if they were indeed connected all along. This capturing and rearranging of realities to present new realities – moving beyond reality, yet retaining something of its verisimilitude, so as to comment upon it – is what I call art (others do as well; see Levinson, 1977b, for more). This third stage of technological development dialectically combines aspects of the first two, but is different from both: art (of which I see pornography as a subset) can be fanciful like the toy but far more serious; and although it can be serious, it is obviously also a big step out of the workaday world.

McLuhan, after Pound, saw artists as "antennae" of the species (1964, p. xi), and art as a crucial component in our understanding of media and their impact upon us. We have already encountered some of McLuhan's insights about aesthetics in his distinction between light-through and light-on (see Chapter 8). In the next two chapters, we bring art center stage.

12

BEAUTY MACHINES

When McLuhan (1964, p. ix) observed that "the machine turned Nature into an art form," he was referring to more than flower prints hung in foyers and living rooms. He was identifying a fundamental consequence of the competition between media: to wit, new technologies do not so much bury their predecessors as bump them upstairs to a position from which they can be admired, if no longer used.

In the case of nature, our involvement with it in the pre-industrial era was as a partner – cooperative, one hoped, and sometimes bountifully rewarding, but also never far from turning barren and even deadly. Our hands touched the soil to coax crops that we needed for sustenance rather than salad; the sea was a locus of transport and food far more than summer vacation; horses were ridden for business and war more often than parades. As we increased our control over nature and our distance from it via technology – as the machine replaced nature as our proximate partner – we gained the ability (or the capacity we already had was greatly enhanced) to appreciate nature for its own sake. Its beauty, its intricacies, the marvel of its workings, all became subjects of art and grand scientific theory (which, on this reading, is a form of ideational art – consider the difference, for example, between Darwin's theory of evolution and practical knowledge about how to breed plants and animals from which Darwin's theory, in part, sprang). The ecology movement – initiated by

Rachel Carson's publication of *Silent Spring* in 1962 – was seen by McLuhan as a logical consequence of Sputnik, and its positioning of humanity for the first time in a position beyond Earth in 1957 (McLuhan & Powers, 1989, pp. 97–8; see also Chapter 5 in the present book).

The implications of such superseded, outmoded, or encircled technologies go deeper than their provision of raw material for art. As a counterpoint and saving grace in McLuhan's view that the most significant effects of media are hidden, he saw them shifting into the limelight, and suddenly becoming amenable to our scrutiny and appreciation, when they were taken off the assembly line of mainstream usage by newer media. Indeed, the replacement by no means had to be complete for the uncloaking of the older medium to occur. Thus, McLuhan saw the narrative structure of the novel as becoming easily comprehensible only after its absorption by motion pictures, whose syntax in turn began to be studied in universities around the world after the advent of television and its adoption of film as part of its content (McLuhan, 1964, pp. ix, 32) (the New School offered a course on film criticism as early as the 1920s, but its introduction into widespread undergraduate curricula awaited the 1950s and 60s). Typically, McLuhan is making more than one point here, which might appear to court contradiction – if novels have become the content of film, and film the content of television, then novels and film seem hardly outmoded or replaced. But McLuhan is actually employing a refined concept of "replaced" – a medium becomes art, and/or content, when it is replaced, not necessarily in its entirety, but in its peak usage, in its performance as the medium or spirit of an age, if we may borrow a page from Hegel. There is no doubt that many more people now watch movies than read novels, and many more than that watch TV.

The question for the digital age in view of this, then, is to what extent its new media have become our work horses, and therein begun to lay bare the workings of older media such as television. In other words, in terms of our understanding of media, the biggest contribution of the personal computer revolution and the Internet may be the light they shed on television as they render it, incredible as it may seem to our television-age sensibilities, into an art form. On this reading of the media, we would have to wait for the next media revolution to gain a more complete understanding and appreciation of the Web and its tributaries.

To better comprehend what may be happening to television before our very eyes, we begin with a consideration of two technologies that have in the past century become art forms.

OF DELICATESSEN AND CONVERTIBLES

One of the joys of engaging McLuhan is coming up with your own examples of the effects he describes. I've eaten a lot of delicatessen food, rarely ridden in convertibles (and have never driven one), but for some reason both occurred to me as prime examples the first time I encountered McLuhan on outmoded technologies as art forms. Both also contain important lessons for the fate of television in the digital age.

The curing of meats – first with salt, then with more sophisticated compounds such as nitrates and nitrites – was initially a technique of preservation. Although ice of course was also available to retard spoilage (which would later be discovered to be caused by bacteria), ice had the unhappy property of melting, which made it unsuitable for long-term preservation unless there was a constant supply. Curing was thus in many cases a better solution. And it also tasted good.

Eventually, refrigeration came to the rescue. Principles of mechanical refrigeration – absorption of heat when a liquid turns to gas will lower the temperature of the surrounding environment – were understood and implemented in the first half of the nineteenth century. By the end of that century, the harnessing of electricity to the task made refrigeration far more efficient; salami, ham, and hot dogs were ready to assume their role as art forms.

What this means is that these meats became primarily appreciated not for their function or durability but their taste. The practical function remained – ham generally lasts longer in the refrigerator than even cooked uncured meat – but that function was no longer delicatessen's raison d'être.

And something else of significance occurred. When the choice was between preservatives and starvation, any ill effects of the preservatives short of causing immediate or demonstrable death were unimportant. The early technology was a trade-off which our ancestors had not much choice but to accept. But once the ingestion of ham and hot dogs became discretionary, the ill effects of nitrites and nitrates became fair and logical game for our concern. Where is the logic of perhaps risking cancer long-term or even high blood pressure just for the taste? The spectre of that logic probably resulted in as many people eating the forbidden fruit as not – for activities that entail risk can be intrinsically attractive (smoking being an unfortunate example) – but the larger point is that the risk as well as the taste now became part of the delicatessen platter.

The lesson applies to all media, and is fundamental to McLuhan's "environmentalism" or Gestaltism: media do not so much have consequences in and of themselves; rather, they have consequences as they perform against the backdrop of a larger environment. This means that changes in the environment, even absent changes in the specific media, will result in changes of performance, and certainly the public perception of that performance. (We saw this in Chapter 9, and its discussion of hot and cool.) Delicatessen not only becomes an art form courtesy of refrigeration, but as our understanding of human health increases on quite separate tracks, delicatessen becomes something of a risk as well. (McLuhan would frequently remark to me that "retrieval" of a medium or a way of doing things from the past could have "deadly" consequences for the present, when our environment no longer knows what to do with it; see Chapter 15 for more on this Platonic thread in McLuhan.) The surprising import of this for television, which we will return to, is that it may be more of a distraction in the next century, when it competes with the Web for screen time, than it ever was as a purported usurper of reading time for books in the century now ending.

The convertible offers other insights into the shifting roles of media. The tops of automobiles were originally rolled down as a way of being cool – a quite practical function in the heat and humidity of summers. By the 1960s, air conditioning (another form of refrigeration) was available in many cars, and was superior for the job, since it cooled without the risk of soot in one's face, or wind-blown hair. For a while, car manufacturers stopped producing convertibles altogether. But the open car returned in increasing numbers in ensuing decades, with a newly "artistic" purpose: in a world of air-conditioned cars in the 1980s and after, one drove in a convertible in order to be cool – not in the sense of being freed from oppressive heat, but in the wholly artistic sense of "being cool, man."

Perhaps, as we enter the new millennium, the operative word could be rendered more appropriately and descriptively as "kewl." But we see in the transformation of the convertible from practical cool to artistic kewl – or being cool in weather to being cool in society – a delightful example of a profound change in function for a given technology, with the linguistic marker for that function staying the same owing to a happy coincidence in the language. Words, in other words, precisely because they admit to and even excel in multiple meanings, are no reliable indication of whether the functions they describe are practical or artistic – whether the technology is at work or has passed into a kind of symbolic afterlife.

Perhaps, in the near future, "watching" television will have a meaning equivalently different yet related to what it is today.

TELEVISION AS ART IN THE DIGITAL AGE

Television, lambasted almost since its inception (e.g., Schreiber, 1953) as an activity even lower than play – a cheap joy, even drug-like (e.g., Wynn, 1977) in its allegedly addictive, destructive effect on us, our children, and some of the finer things in culture – is already experiencing some of its first ascensions to art. Not so much because of computers (as yet), but as the ubiquity of movies on cable and via video rental have broken the once near-monopoly of network programming on television, the motifs and even techniques of TV's "golden age" are given increasingly higher regard. The broadcast of live dramas on CBS-TV's *Playhouse 90* (1956–61; see Brooks & Marsh, 1979, pp. 498–9) are now recalled in praise accorded the "legitimate" theater (which became "legitimate" in proportion to the rise of motion pictures). Even slapstick situation comedies such as *I Love Lucy* and *The Honeymooners* are not only rebroadcast, but often with sotto voce commentaries on their comedic artistic significance (as, for example, on the *Nick At Night* cable TV programs in the U.S.). The cable channel "TV Land" also replays commercials first broadcast in the 1950s and 60s, purely for the audience's pleasure of seeing them again.

Moreover, in a self-reflexive twist that demonstrates how figures and grounds in media evolution keep changing places, motion pictures ranging from *The Fugitive* (1993) and *Mission Impossible* (1996) to *Leave it to Beaver* (1997) and *Dennis the Menace* (1993) have been bringing to the big screen a parade of characters who first attained success on the small screens of television. Having pushed motion pictures into higher esteem by replacing it as the work-a-day audio-visual medium in the 1950s, television now provides an additional service to the movies on the other end of the curve by becoming its content. This "reversal" of figure/ground, as McLuhan would call it (see Chapter 15 in this book), is almost perfect: TV, which commences its reign as the medium of the postwar age by making film its content and art, now becomes the content and art of film in the new digital age.

The reason, of course, is that the digital age – including not only computers but cable television and video rentals – has changed the nature of film and television, or at least the way they are seen, and hence their role in our lives. Thus, the presentations of movies on cable and video, not movies per se, are the real instigators of the change in the status of TV; and computers, not motion pictures, will finish that job. The figure/ground shift is much more than a simple reversion – motion pictures do not become what TV was prior to the shift, nor do motion pictures recapture the position and role they had prior to the advent of TV. Rather, classic network television, because of the advent of new media, now becomes appealing as content and art, and joins the novel in that function for motion pictures. New media, then, not only spearhead the evolution of society by what they accomplish in their own rights, but by the ripple effects they exert on the changing performances of earlier media. Reversals in such performances – as in the case of television and motion pictures – are thus progressive, or indicative of genuinely new developments, rather than merely recapitulating what already was.

Yet the outmoded medium newly discovered as an art form need not necessarily change in any way other than in our perception of its role – nature, after all, remained nature, even after the machine put it on the stage. What, then, are we now beginning to see in television that makes it interesting to us as art, or as the content of motion pictures? The answer is that it is some of the same of what television always did – but, more particularly, the aspects of TV that appeal to us most as art are those aspects that contrast most with the performance of equivalent newer media. If cable television and video rentals are making stand-alone, individual movies available to us on the television screen as never before, what could be more intriguing as an "aesthetic" of a recently bygone age than characters from continuing weekly television series? If the Web excels in the accidental hypertext connection, how could the carefully plotted performance of a *Playhouse 90* not loom large in nostalgic cachet?

Significantly, although the "liveness" of *Playhouse 90* and other early television – what in today's parlance would be termed "realtime" performance – is admired and even cherished as one of the distinguishing features of "classic" TV, television when archived on videotape or the Web becomes the opposite of live or realtime. Not only are the performances not actually occurring when we view them – having ipso facto been recorded earlier if they are the content of, say, a videotape – but neither are they seen at the same time by everyone. Thus, in contrast to the classic television audience, comprised of millions of people watching the same program at the same time, the viewers of tapes on VCRs and snippets of TV on the Web are individuals who partake of this new form of television at times of their own choosing.

The classic simultaneous TV audience of course still convenes – and now, via satellite, on a truly immediate international basis – for events such as the Superbowl and the Academy Awards. The global audience was assembled for the wedding of Charles and Diana in 1982 and then again for Diana's funeral in 1997. Such classic TV audiences are the epitome of an evolution of simultaneous audience that begins in public squares and theaters, moves into movie theaters with the Lumières and their projection of motion pictures on to screens, and finally achieves national and international coherence – a single Academy Awards speech, to everyone in the world at once – in the twentieth century through radio and television. But television, as the content of individual VCRs, reverses the process, and moves the motion picture back from the public screen to the private kinetoscope, whose movies could be seen by just one person at one time.

Again, however, such reversal is far more than just recapitulation. Whereas viewers of kinetoscopes were obliged to go out of their homes, to arcade-like parlors open at given hours, to see *Fred Ott's Sneeze* and other favorite flicks, the viewers of a videotape on a VCR can indulge their inclinations any time they choose. In that sense, television viewed on pre-recorded videotape has more in common with the book than the kinetoscope.

Television archived and available in "RealVideo" on the Web is even less tied to time. In order to read a book or see a videotape, one must at some prior point make a decision to buy or borrow the book, or buy, borrow, rent, or record the videotape. If one desires to read a book or see a videotape, and that prior specific action has not been taken, then the desire will be unrequited. In contrast, anything available to the public on the Web is there for the taking twenty-four hours a day, with no prior decision needed, except to arrange for general Internet access in the first place, via the requisite personal computer equipment (of course, Internet connections like everything in technology and life are fallible, so there are no guarantees as to instant availability).

This brings us to a consideration of the machine – the Internet – in the transformation of television into an art form, and its reversing of relationships among prior media. For it is the contour of the Internet, what exactly it accomplishes and how, that molds the art in television. Or, we might say that the specific shadows cast upon television by the Internet bring out the art we now see on the older screen.

Before turning more fully to the Internet's part in this transformation, however, we need to look at another aspect of television, unique to its role for the past half-century of viewers, which has an impact on the art form it is becoming.

FIRST LOVES AND NOSTALGIA VERSUS ART

McLuhan's observation that "the machine turned Nature into an art form" (1964, p. ix) is, on closer contemplation, not very typical of media and their impact in general on one crucial account: nature is by no means your typical medium or environment. It is, rather, a "medium" or environment that transcends its transformations – including by the Industrial Revolution – and can be discovered anew, in almost pristine form, every time it rains or snows or shines.

In contrast, all artificial media have discernible points of beginning (although this is of course less clear with ancient media such as the alphabet) and transformation into art forms if and when that time comes. It is hard to imagine anyone looking at fine handwriting as a crucial practical skill in an age of word processing, or even typewriting; instead, fine handwriting has become ceremonial calligraphy. But we can still feel and fathom the practical advantages of a sunny day, as well as its beauty, even if the industrial and now the information revolution have long removed us from the farm.

Because nature never goes totally out of use – or out of style – it is not subject to the same powerful tugs of nostalgia we feel for media that we grew up with, became accustomed to, and may now find suddenly replaced in some functions by new media. Such nostalgia tends to coincide with the treatment of the older medium as art when it is first displaced – we are both nostalgic for the golden age of television and may regard it as an art – but, eventually, the nostalgia wears off (as consumers of the medium in its heyday leave the scene), and the medium survives as an art form to the extent that it has qualities that humans find attractive whether or not they have had previous experience with the medium. No one who today enjoys delicatessen does so in fond remembrance of the time when lack of refrigeration required those meats to be so cured for preservation; rather, delicatessen like spice is today enjoyed because of its taste. This suggests that we may yet be a half century or more ahead of the time that television finds its "objective" artistic taste – that is, a taste in the public not imbued with nostalgia for the days in which there were no personal computers, no Web, and television was the coolest cutting-edge medium in town.

Closely related to nostalgia is an effect I call the "first love" syndrome:

we never get entirely over the media which accompanied us as we first came of age socially, in business, or which we first mastered as a means of accomplishing a specified kind of task. These first loves are almost ever and anon the media we feel most comfortable with, or adopt as our standard of performance for competing media of similar function (see Levinson, 1997b, p. 166, and Levinson, 1998a, for more). I saw this effect most frequently in the quick turnover of word processing, telecommunication, and data-base programs in the 1980s. Once someone learned how to write and print out with WordStar, Word Perfect, or one of the many other programs, no other word processor would feel quite right, and usually only the purchase of a new computer (for other reasons), which had diminished capacity to run the initial program, would result in the first program's abandonment. It took me until 1988 to forsake my CP/M machines for DOS, and I still prefer DOS to Windows for many tasks. And, yes, I'm writing this right now in WordStar. But I try not to mistake my comfort with these programs as evidence or proof that they are objectively better than their newer rivals (for more on the continuing attachment of some writers to WordStar, see Sawyer, 1990/1996).

Having grown up in a world in which cultural imprimaturs were bestowed by appearances on the *Ed Sullivan Show*, on the *Tonight Show* with Johnny Carson, on *Nightline*, my guess is that appearances on the Web – certainly via text in publicly announced chats – will never have the same import in my eyes. Similarly, whatever the merits of publishing and reading online and/or on a screen, my fundamental internal sense of where text should be is on paper – whether newspaper, magazine, or book. I expect to see authors' names on packages of pages in bookstores and libraries; I come across writers every day on the Web, but I have been conditioned to look for ultimate confirmation of their status on the shelves.

That the Internet is contesting the positions of both television and print is indicative of the scope of its surge into the mainstream. Television, despite decades of apocalyptic pronouncements to the contrary, never posed that sort of challenge to print, offering at most a distraction from rather than a replacement for the reading of words affixed to paper. With television and books and newspapers now all under pressure from the Web – all first loves of ours in their different functions – they may be in for a very special send-off, unique in the life cycle of a medium, as they all sail off in the same slow boat into the sunset of art in the twenty-first century.

That assumes, of course, more or less equal triumph of the Internet over each of those media. Since the Internet competes with television for the same screen – literally, in the case of Web TV – television may be the easiest conquest. An anonymous reference to television as "for the Web-impaired" that I encountered on the Internet may be the handwriting on the wallscreen.

Print on paper may have far longer endurance, since it comes with batteries included – requires no source of energy to operate other than what is already present in the environment – and supplies reliable locations for favorite texts, in contrast to screens with texts that constantly change (see Levinson, 1998a, for more on the survival prospects for books in the digital age). Physical places – even for words on pages – have not only aesthetic but enhanced practical appeal as a specialty in a discarnate age.

By the same reasoning – television and the Internet already both employ screens, books do not – we can expect television to be more of a distraction than are books to work and fun online, as discussed above, as TV puts up a rear-guard action to hold on to its dwindling status as a mainstream medium.

What will the Internet look like – what will it do for us – as it assumes that predominant position in society?

THE INTERNET AS MAINSTREAM

We have already considered what life is like when we work on the Web (see Chapter 11). Here we consider, in a sense, the other side of the same online coin: what will the world be like when the Internet works for us as the pre-eminent ratifier of culture and reality, when its displacement of television in that role is as complete as television's has been of radio for the past five decades?

To appreciate the enormity of such a new displacement, think about Ronald Reagan, "the great communicator," on television, and how comfortable people in America were with that performance; now picture Reagan in a live online chat room with the American people, engaging in a written dialog with his questioners. If the gulf between radio (which went on after its displacement to thrive as a vehicle of rock 'n' roll) and its successor television was large – epitomized in the political realm by Franklin Delano Roosevelt (radio) versus Ronald Reagan (TV) – the gulf between staged, one-way television and the interactive, text-driven Web seems almost unbreachable: Reagan is simply unimaginable in a serious online dialog of text. (On this reading, John F. Kennedy, the first and in many ways still the most successful televised President, retains some of the drive and heat of radio and Roosevelt. He thus stands midway on the radio/television continuum between Roosevelt and Reagan – with Nixon closer to Roosevelt on this scale. Kennedy, moreover, prized not only orality and visuality but literacy, and likely would have done quite well online.)

Of course online chats, especially if we are looking towards the future, need not be conducted via writing. RealAudio and RealVideo on the Web are already presenting interviews and speeches online in the manner of classic radio and television, as we have noted throughout this book. But as we have also seen, the environment in which such presentations are situated – on a personal computer with a Web browser on a desk in contrast to a radio in an automobile or kitchen or bathroom or a TV in the living room or bedroom – can make all the difference. The very fact that word processing and so much else of what is done on the computer entails text and occurs on the same machine as RealAudio and RealVideo lends an expectation of text even to these non-textual performances. Most of my interviews on RealAudio provided occasions for listeners to ask me questions via live text chats; RealAudio on the Internet thus made them writers as well as listeners.

Radio and television are transformed in another profound way on the Internet, also mentioned earlier. To be present on the Web is ipso facto to be archived on the Web: realtime online is by virtue of being online also a readily accessible past time – indeed, so fundamental to the Internet, and even computing itself, is this storage capacity, that deliberate decisions not to archive are usually required for it not to happen. I noticed the advantages of such default archiving in the very first online course I taught for the Western Behavioral Sciences Institute in 1984. The course was a month in length, and every comment made by the students and by me, from the moment the course began, was available for recall and scrutiny throughout the month. I call this the "intellectual safety net" of online education.

When Presidential addresses and newscasts are presented online, that safety net – that retrievability and accountability – will pertain to them. As a counterpoint to the spontaneity that the ease of communicating online provides – in contrast to the careful scripting and production of television – we will also find an endurance of information online that far exceeds not only that of broadcast media, but perhaps print as well. Certainly the dissemination of information online is far faster and wider than print on paper. It is already far more durable than print in newspapers or magazines, and there is no reason to think it will not last as long or longer than books. When we take into account the ease of production and accessibility on the Internet, along with its capacity for dissemination and retrieval, the magnitude of its differences with all prior media becomes clear: none of these media – not television, not radio, not books – offers an equivalent package of opportunities and impact. On the one hand, it is almost as easy to write on the Web as it is to speak. On the other hand, anything spoken on the Web is in principle spoken forever.

A logical outcome of such increased retrievability is that more care will be taken with the initial presentations. After all, if I know that every word which I write in an online course will be available for perusal and study of students throughout the course, I will likely accord somewhat greater attention to their composition than the words I speak in in-person classroom discussions. This suggests that the very production and presentation of work online may serve to improve its quality. Taken to its conclusion, this surprising result holds forth the possibility that work conducted online may be of superior quality to equivalent work conducted offline – and to the extent that work of all kinds is increasingly done online, the prospect that the Internet may be an occasion for the improvement of work in general.

The result is surprising because we have already seen, in the previous chapter, the degree to which the Internet blurs the distinction between work and play. Work as play (and play as work) hardly seems the most conducive to improvement or perfection of work for the posterity of the archive. But what McLuhan had in mind when he talked about the proximity of formerly disparate functions in the new electronic milieu was something which went beyond the blurring of just any two distinctions. Rather, he was thinking of a much more general melting pot, that encompassed not only the spontaneity of play and the seriousness of work – but the perfection of art. If the Industrial Revolution had severed childhood from adulthood, it also had squeezed out the personal perfection of the handicraft from the mass production of the machine.

That idea – that the machine steps on the aesthetic, and thereby the artist – has of course itself been around since at least the age of Romanticism, and its first protest of the Industrial Revolution, now nearly two centuries ago. McLuhan's notion that electronic media reverse that process by reintegrating aspects of art into everyday life was something new. It arose, as did most of McLuhan's ideas about electronic media, from his view that the very process by which we perceive television – all-at-once, all-around, involved, in contrast to the page-by-page specificity and distinctions and inherent aloofness of print – was an integrating, artistic mode of experience.

In the next chapter, we explore the extent to which this vision may be achieving literal fulfillment in the work we conduct online.

13

BALINESE AT WORK ONLINE

McLuhan was fond of quoting the Balinese saying, "We have no art, we do everything well"; in fact, it appears in at least six of his books (McLuhan, 1964, p. 72; McLuhan & Fiore, 1967, p. 137; McLuhan & Parker, 1968, p. 6; McLuhan, 1970, p. 312; McLuhan & Watson, 1970, pp. 118–19; McLuhan & Powers, 1989, p. 15). Earlier, McLuhan (1962, p. 85) had noted also about the Aivilik that "they have no word for art," and quoted Edmund Carpenter's observation (in *Eskimo*, written in 1960) that "every Aivilik adult is an accomplished ivory carver."

It is easy to imagine doing everything well online. Virtual constructs are easier to manipulate than physical objects. Human relationships are safer as well as easier to pursue online. And the information needed for preparation and completion of most tasks is usually much closer at hand online than in any physical surrounding, except perhaps the biggest research libraries in the world.

Does this, then, mean that everything is indeed done well online? Can we do as much on our non-ivory computer keys as the Aivilik with their ivory?

In pre-industrial societies such as the Balinese and Aivilik, the resource that most allows them to do everything well – or at least, offer a philosophy that they do – is time. As the clichéd but not entirely untrue story of our modern age and its discontents has it (clichés should, after all, have some truth in them – that is why they became clichés in the first place), we pay for the mass availability of commodities with a sharp reduction in the amount of time and attention accorded the production of each. Ford's assembly-line motor cars, whatever their other huge advantages over horse-drawn vehicles, could not possibly reflect the amount of individual care given to the raising of each horse and the construction of each carriage.

The information age, successor to the industrial age, moves its virtual products far more rapidly than the assembly line. Little or no time is spent in the automatic linking of text to text on the Web, or in transmission of e-mail – expenditure of time, when it does occur, is seen as poor performance, not a benefit, of the system. And, indeed, little or no attention is accorded the performance of the links and the e-mail, except when they break down. And yet, we have the sense that the era of the Internet has furnished us with more time to attend to our work, at least when that work is performed online. Where in the instant world of the Internet does this time originate, and where in our work is it brought to bear?

Time is a most fascinating multiple-edged sword, to say the least. The capacity to conduct some tasks at speed of light frees up time to conduct other tasks, or other parts of the same task, with more leisure and presumably more care. Although the assembly line greatly sped up production relative to the handcrafted age, it did not speed up production quite enough to give workers the dividend of time to do some tasks extremely well, or give some tasks the personal touch. Thus, McLuhan, although he was not referring to the Internet, was quite right that the sheer speed of electronic communication retrieves the Balinese attention to individual quality and detail.

This attention is implemented, online, not in the operation of hypertext and e-mail but in their inception – on the occasion the e-mail is written or hypertext link is first devised. But did not Henry Ford and indeed all manufacturers make use of inventions on which considerable time had been lavished in their inception, apropos Thomas Edison's famous observation that invention is one percent inspiration and ninety-nine percent perspiration?

Yes. But this was for the inception of the invention, not its products.

And the Internet has another consequence, discussed at length throughout this book, that makes its world very different from the Age of Invention and Edison: the Internet, by allowing anyone and everyone with a personal computer to come to the party, affords attention to inception and careful detail to a multitude of potential Edisons. If the Internet allows us to do everything well, it will be not just because of the time for careful work that its speed sets free, but also because the "us" in that hopeful equation is a vastly bigger group of minds than it ever was in the past.

Of course, it is far too soon to empirically demonstrate or refute the proposition that the Internet will be bringing up a bevy of Edisons – of Morses, Daguerres, Fox-Talbots, Bells, Lumières, Marconis – not to mention the inventors of wondrous things other than media in the now bygone Age of Invention. The environments of those geniuses likely had ingredients unrelated to the Internet and its liberation of time, aspects which would need to be present, at least in equivalent form, if we are to see a new age of invention in the twenty-first century on a par with that of the nineteenth century which brought us so much of the world in which we have lived and worked in the twentieth century.

We proceed in the following discussion, then, with the proviso that factors beyond the Internet's capacity may be necessary for it to help us do everything well. But that still leaves us lots of territory to explore in determining how the Internet is helping us move in the better, Balinese direction.

We begin with McLuhan's description of electronic environments as "mythic."

MYTHIC PROPORTIONS AND HERCULEAN TASKS: VERTICAL VS. HORIZONTAL

When McLuhan (e.g., Carpenter & McLuhan, 1960, p. ix) says that the electronic age is "mythic," he does not mean myth in the sense of falsehood or misinformation. Indeed, to the contrary, he is using myth in its much older, original sense of conveying information on a richer level, one which resonates (a favorite McLuhan term, because it is an acoustic metaphor) with fundamental truths that may not be as available to commonplace, work-a-day scrutiny. McLuhan's use of myth is thus much like Joseph Campbell's, recently employed by Sylvia Engdahl (1990) in her work on "space-age mythology," which she sees not as tall-tales about space but expressions of our inherent citizenship in the cosmos. It is in this sense that electronic "myth" is a piece with the Balinese conception of art and doing everything well.

How does the Internet figure in this?

Time turns out to be a crucial element in the creation of myth. In classic myths – I like to call these "vertical" – hundreds sometimes thousands of years are needed for the requisite number of minds to process the retelling of an event to transform it from recollection to history to myth. One consequence of electronic media – manifest by radio and television in their broadcasts to immense simultaneous audiences, and attained more or less also in the conglomerate mass audiences of movie theaters (not literally assembled all at one time across a nation, but achieving much of that effect over a period of weeks or months) – is that the requisite number of minds to generate myth can be assembled almost instantly, or certainly much faster than classically. I call these modern, electronic myths "horizontal."

Indeed, the speed-up in attaining the requisite number of minds for production of myth begins with the printing press. Columbus's voyages to the New World achieved mythic proportions – again, in the positive sense of inspiring people to great actions, because it captured a fundamental yearning of humanity and writ it large – via published reports that flowed from the presses of Europe in the 1490s (see Levinson, 1997b, pp. 25–8, for details). Significantly, the oral sagas that described Leif Ericson's voyages to North America half a millennium earlier had no equivalent effect. They were anecdotal – literally (from the Greek root), unpublished. The area of their mythic influence was too limited – Norse culture – and not enough time had elapsed for the sagas to percolate through enough minds to achieve full-blown mythic status the old-fashioned, vertical way. Thus, no Age of Discovery, no revolution in European political, social, and economic systems, resulted from the Norse trips to America.

A good example of the confluence of mass media – printing press, motion pictures, radio and TV – as well as the Internet in investing a real event with mythic proportions in the twentieth century is the media coverage of the sinking of the *Titanic* and its consequent ascension into popular culture. The raw materials were of course there to begin with: the biggest liner ever built, thought by some to be unsinkable, founders on an iceberg and slips into the freezing waters of the North Atlantic on its maiden voyage with the loss of some 1,500 passengers. Further, these passengers included some of the richest people in the world at that time – Benjamin Guggenheim, John Jacob Astor, Isidor and Ida Straus – alongside people from all other walks of life. The ship went down with its Captain and its builder; the head of the company (the White Star Line) that owned the ship survived via one of the lifeboats, which managed to save 709 people. A nearby ship, the *Californian*, might have saved many more, but failed to respond to *Titanic*'s desperate calls for help via Marconi radio, light (flash) telegraph, and distress rockets.

That last aspect of the *Titanic*'s sinking is itself a tragedy of telecommunications of mythic proportion. And so is the story of overconfidence in the *Titanic*'s ability as an epic tale of Greek hubris. The self-sacrifice of some of the wealthiest passengers, the selfishness of others, the conflict between the lower classes and authorities on the ship in its final minutes, all touch on themes and nerves worthy of treatment by Homer and Shakespeare.

The mass media of the day – newspapers – were quick to oblige. By the end of the twentieth century, more than 35 movies and 100 books had told and retold the *Titanic*'s story (see *Titanic: Secrets Revealed*, 1998, and Heyer, 1995, for details). So prominent was the disaster in the public's mind in the 1980s that Robert Ballard was inspired to finally locate *Titanic* two miles under water in the North Atlantic, after numerous attempts by others had failed. This dose of history-become-real inspired a new series of media presentations, including the 1997 movie *Titanic*, the biggest earning movie of all time, at the time of this writing in 1998.

Unsurprisingly, the Internet hummed with Web pages about *Titanic* – the ship and the movie and the ship in the movie, all the same in myth-land – throughout 1998. And newspapers, radio, TV, and new books of course joined the myth relaunchings. Thus, the passage of *Titanic* into a nobler world than just death at the bottom of the sea was occasioned by an incredible wave of various media and reality, each feeding off the other and stimulating both further myth and greater understanding of historical reality.

That last effect – fuller and more accurate understanding of reality – is why I hold such horizontal myths to be profoundly beneficial, examples en masse of the media helping us do more things well. Vertical myths usually achieve their status far too late to result in any real increase in our understanding of history. I suppose it is possible that someone moved by the many myths of King Arthur and the Knights of the Round Table might go on to discover some actual, heretofore unknown historical site related to Arthur – just as Heinrich Schliemann identified Hissarlik as Homer's Troy – but such discoveries, owing to the sheer age of their objects, will almost always be far more tentative and thus historically less conclusive than Ballard's location of *Titanic*.

Such horizontal mythic effects happen on television, and other mass media, for a myriad of smaller events all of the time. Critics of mass media miss the fact that, by any reasonable reckoning, more people are better informed now than ever before. Undoubtedly, some of the information, much of it in some cases, is distorted or outrightly in error. False stories deliberately leaked to the press can result in myth in the sense of deception. But if we agree with Jefferson that the best remedy to misinformation is more information – because people have the rational capacity to separate truth from falsity, given the presence of sufficient true as well as false information – then we can expect the massive dissemination of information by and large to create myths that are closer not further from reality. Thus, the initial myth of Kennedy and Camelot has been tempered a bit by decades of reportage on Kennedy's extramarital affairs; whereas the initial myth of Clinton as sexual philanderer was at least partly substantiated by the report of the Independent Counsel sent to the U.S. Congress in September 1998, and by Presidential confessions and apologies. In both cases, the continued flow of information brought the myth into closer consonance with reality. (This is not to say that I think that the sexual activities of public officials are appropriate subjects of political retribution – as I argue in "Can Only Angels Be President?" in Levinson, 1992, pp. 151–3, I do not. But the above examples nonetheless support the Jeffersonian point that the continued flow of information tends to bring us closer to truth than fiction, whatever the subject.)

Further, the Web sharpens this process for us as individuals by allowing each of us to pursue more information about the myth in our own way, at times of our choosing. Again, there are no guarantees that information we may find on a Web page is truthful – any more than there are guarantees that the information presented to us by the gatekept media of newspapers and television is true. But the same Jeffersonian principle applies to the Web, and even more so, because the Web is comprised of a multitude of authors/publishers rather than an oligarchy as in the mass media. Unless every single Web page on a given subject is tainted with the same misinformation, we are likely sooner or later in our extensive browsings on the Web to come across information that exposes the deceptive myth.

Does this media propensity for the creation of truthful myths, and the contribution of the Internet in particular as a myth-correction mechanism, mean that these electronic media are transforming us into Balinese who do everything well? No.

But it does seem to be helping us do a part of our life's work better – the part that entails our being informed citizens in this world. If artists are indeed the antennae of our species (McLuhan, 1964, p. xi, quoting Pound), then electronic media surely seem to be providing everyone with a multiplicity of more powerful antennae.

But what of other aspects of our lives?

What impact, for example, have electronic media had on work that has indeed been traditionally identified in our culture as artistic?

THEREMIN AS THERAPY FOR ART

From the point of view of computer storage and transmission of digital data – binary code – there is no qualitative difference between text, images, and music. There may be more or less of the binary code, but it is all the same code. The differences arise in the way the words, pictures, and sounds are entered into the computer system in the first place, and even more so in the way they are transformed from binary data back into stuff we can see and/or hear. Thus, keyboards, visual scanners, and MIDIs (Musical Instrument Digital Interface) enter respective texts, images, and music as binary data for storage and/or transmission, as on a Web page. That data is eventually transformed back into texts and images on screens, and music through speakers.

The discontinuity in the storage/transmission of the data and the way it originates and ends – the universality of the language of storage/transmission (a common binary code for everything) versus the particularity of its inputs/outputs (reading, seeing, hearing) – stems from the fact that the process of storage/transmission, binary code, is not directly intelligible to our senses and cognition. It works so well as a conduit of communication precisely because it has no literal resemblance to what it is communicating. The alphabet may have been the first deliberate medium to implement this strategy – it communicated to many more people than did hieroglyphics, and apparently a far wider range of concepts (see Levinson, 1997b, Chapter 2) – and DNA can be considered a natural "digital" medium in this regard, in that its instructions for the organization of proteins look nothing like the organisms and organs they bring into being.

This discontinuity is the basis of some analogic communication as well. The grooves on a phonograph cylinder or record may have highs and lows equivalent to the highs and lows of the sounds they represent, but placing one's ear to the recording yields no sound. Similarly, the electrical patterns in telephone wires may be analogous to the sounds they represent, but cutting open a telephone wire and sticking it in one's ear is a useless method of eavesdropping – one might as well go out in the street and hope to hear radio transmitted over the air, *sans* radio, which, come to think of it, is an even more vivid example of the utter difference between the input/output (sound) and its mode of transmission (in the case of broadcasting, analogic electronic patterns that resemble the sound waves but are not themselves sound). In all three of these cases – phonograph, telephone, radio – our ears are simply deaf to the mode of storage or transmission.

Shannon and Weaver (1949) called this process "encoding" and "decoding" – encoding an energy form, perceivable by humans, into a form that is not, for the purpose of storage or transmission, and then decoding it back to its original form after the storage or transmission – and described it as the basis of all communication. Prior to digital computers, the encoded form differed from medium to medium – grooves in a record versus patterns of electricity in phone wires, etc. The digital improvement in this regard, then, was to make the encoding process the same for all media.

And this has begun to open some interesting possibilities for art.

For if the keyboard can be used to enter text, which has the same digital form as music stored as digital data (as in CDs), the keyboard can also be used to enter digital data that will play on speakers as music. In other words – literally and figuratively – the origin of music, as long as it is digitally encoded, need not be wedded to a prowess with string or woodwind, not even a talent for the keyboard of a piano or organ. All that is needed is the idea of what the music should sound like, and the knowledge to write a routine for production of that music on a computer keyboard. Indeed, that routine need not even be typed into a computer – it could, in principle, be spoken.

The electronic separation of prowess with a physical instrument from the initial production of music in effect began in 1919 with the "theremin" invented by Leon Theremin (né Lev Termen – my second Russian namesake from that era with an important role in communications – the other was Lev Kuleshov, who performed the famous montage experiment in the Soviet Union that same year). Musical performances had of course been separated from their instruments ever since Edison accidentally invented the phonograph in 1877. But the theremin was different in that it produced a kind of music without instruments in the first place. It was a creator, not a reproducer, of music.

Of course, the theremin does not operate via telekinesis – the direct mental control of material objects that we find in science fiction. But its control of pitch and volume via the operator's movement of hands and body near its two metal antennae – which vary the frequency of two sound oscillators – provides a pretty good approximation of mind literally over material instrumentation, without the mediation of lips and fingers on the instrument. Indeed, if we think of the conductor waving instruction to the orchestra, and the thereminist waving instruction to the theremin, Termen's invention can be seen as a primitive artificially intelligent orchestra – a machine that has the "intelligence" to take cues from the gestures of its conductor, just like the human members of an orchestra.

The trembling tremolos of the theremin – which do sound as if they come straight from some feverish brain – were quick to capture the public's imagination and approval. Termen left the Soviet Union and came to America in the 1920s; his theremin was heard at Carnegie Hall, in Leopold Stokowski's orchestra, and other concerts reviewed on the front pages of *The New York Times* (see Steven M. Martin's 1995 documentary, *Theremin: An Electronic Odyssey* for more). Isaac Asimov likely had the theremin in mind when he wrote in his award-winning science fiction "Foundation" series how the Mule controlled the loyalties of his victims by mental telepathy: "To me, men's minds are dials, with pointers that indicate the prevailing emotions....I learned I could reach into those minds and turn the pointer to the spot I wished" (Asimov, 1945, p. 164). Roger Ebert (1995) notes in his review of *Theremin* that the instrument was used "by Hitchcock to suggest mental illness in *Spellbound* [1945]," by Robert Wise "to accompany alien life forms in *The Day the Earth Stood Still* [1953]...by Billy Wilder to suggest alcoholic disorientation in *The Lost Weekend* [1945]...by Jerry Lewis to suggest looniness in *The Delicate Delinquent* [1957]," and by other filmmakers to touch wells of emotion in the minds of viewers. Rock music's biggest groups – the Beatles, the Rolling Stones, the Grateful Dead – would soon pick up the theremin's ethereal sounds, most famously the Beach Boys in their masterpiece, "Good Vibrations," in 1966 (Brian Wilson's production also included such exotic instruments as the jew's harp, held between the teeth when played).

Meanwhile, Robert Moog, inspired by the theremin he built from a hobby kit in the 1950s, went on to invent a new kind of electronic-music instrument, the Moog Synthesizer, in the 1960s. The Moog had a much wider range of sound – and means of controlling and specifying it – than the theremin, and Walter Carlos' virtuoso use of it in his best-selling "Switched-On Bach" album in 1968 installed the instrument at the intersection of classical and avant-garde music for decades. Rock musicians also employed it, as they did the mellotron – a precursor of current digital samplers – which called forth and manipulated sounds from recorded tapes.

Interestingly, both the Moog and the mellotron used keyboards to select and present their sounds, and thus harkened back to a mode of musical creation in which physical contact with an instrument was not as removed as it had been with the theremin. Indeed, Elisha Gray – the hapless independent inventor of a telephone who filed his patent a mere few hours after Bell in 1876 – had attached a keyboard to telegraph transmitters (that produced different pitches) and played an octave in 1874. And Thaddeus Cahill's Telharmonium at the turn of the last century employed two keyboards to produce its electronic sounds. As we will see in more detail in the next chapter, McLuhan's concept of the "rear-view mirror" – our tendency to walk into the future with our eyes fixed on the past, a conceptual nostalgia – explains this attachment to keyboards as input devices for music as well as words. Its role in electronic music was to make most of its instruments not much different from the classical piano, harpsichord, or organ in terms of physical talent required to elicit sounds, especially in live performances – although the non-electronic instruments usually require more dexterity – leaving the theremin as a high water mark in the direct conversion of music heard inside the head to music actually heard in the world.

In what way would such direct translation of virtual to actual, in which music is produced absent physical ministrations, be an improvement?

The discussion in the previous chapter about what I call the "first love" effect – our penchant for taking the media with which we came of age as our standard of excellence (this might be considered a subdivision of McLuhan's rear-view mirror principle) – shows how difficult such questions about improvement in art can be. We grew up in a world in which wonderful music came from people playing pianos, strumming guitars, blowing into saxophones. We link as seemingly inextricable the physical virtuosity in playing these instruments with the music produced. We recognize that the addition of electricity to the process – as in the electrification of guitars – vividly changes the playing and the sounds produced. But if Eric Clapton, the maestro electric guitarist, displays a virtuosity different from that displayed by the master acoustic guitarist Andres Segovia, it is nonetheless as tangible. The electric component in this case transforms but does not eliminate or even lessen the necessity of touch.

With such a legacy in our minds and ears, how can we take seriously an instrument whose pitch is modulated by the wave of a hand up and down near its vertical antenna? Artists may be the antennae of our species, but can such hand-waving near antennae possibly be art?

We can begin to understand how it could be – and indeed, why it may in some respects be an improvement of art – by looking at the constituents of traditional art, in this case, of the guitarist or pianist. The music created usually begins with an idea, a conception of the music, in the performer's head (the exception would be the performer casually playing the instrument, and producing sounds which in turn inspire ideas, which in turn inspire sounds). The idea is implemented, the music comes to life, when the guitar is strummed or picked, or the piano is played. In this fruition of the musical idea, the physical construction and composition of the instrument of course also plays a role. Musical virtuosity in this traditional model comes in three interrelated parts: (a) the quality of the idea, (b) the quality of the performance, (c) the quality of the instrument. The proximate performer usually has exclusive input only into the second component – the actual performance. The idea – the melody and harmony – may have originated with someone else. The instrument was likely conceived and constructed by one or more other persons.

Now let us consider what happens when someone plays a theremin. The process begins the same way – with (a), an idea in the performer's head, either of the performer's or someone else's initial creation. Then hands are waved near the antennae – this constitutes (b), the performance. The physical instrument, (c), is the theremin, and bears the same relationship to (a) and (b) as in the traditional model – it is conceived and constructed independent of the performance.

So where in terms of this model is the difference between the piano and the theremin? Clearly, it resides entirely in (b), the performance. Although skill and training may be needed to elicit specified music from the theremin, we correctly feel that the talent in this performance is less crucial to the outcome than in strumming a guitar or playing a piano. This can lead to a perception of the theremin as a less worthy instrument. "Is the Theremin a serious instrument?" Ebert (1995) asks. "I think not."

But note that although the talent may be diminished in (b) – the performance part of the theremin – it is by no means diminished in the overall production (a, b, c). Rather, it has been redistributed, shifted to (a) and (c) – the idea and the instrument. (Singing, in this account, would comprise yet a different kind of distribution, in which the idea and the performance took precedence over the physical "instrument" – the vocal chords.)

We can discern the greater contribution of the instrument in electronic music more clearly when we consider the mellotron, or any device that generates sounds from pre-recorded tapes (in today's terminology, digital samplers). Although the performer on such instruments exercises talent in the choice of sounds, determining their duration and blend, this talent is not on a level with those who coax sounds out of strings and woodwinds.

The guitarist may say, with charm and McLuhanesque metaphor, that the song is in the instrument (the medium is the message), but surely much less of that song is in the guitar or the clarinet than in the digital sampler in which every elicited sound was created by someone else.

But if the advent of electronic music in the end is but a shift in talent from performer to instrument, with the performer contributing less and the instrument more, how is that an improvement in art? Why are digital samplers, and devices which allow music to be created at computer not piano keyboards and therein shift the responsibility of sound to another machine, in any way better?

One part of the answer is that although they may not be better for the performer, they may be better for the listener. They may in the long run result in music that at least some people, freed of the first-love effect and the rear-view mirror, find more satisfying. Even in the case of the theremin, if listeners find that its music touches their souls in a way that none other does, what does it matter if the music was summoned by a magician waving hands, rather than sliding fingers upon strings and keys?

The other part of the answer may be found in (a), the mental creation of the music – the hearing of it in the performer's head – that precedes the actual performance. By de-coupling this virtual from the ensuing physical performance – or rather, by diminishing the importance of the physical performance as a necessary mediator between the musical idea and its outcome on the instrument – electronic instruments may be opening a musical gateway from the mind whose price of passage was heretofore physical adeptness.

This liberation of the mind from its immediate dependence on tangible things for expression of its creativity is indeed the key to how the digital revolution may be making everything better, and not just in music.

VIRTUAL VIRTUOSITY

Virtual music – whether in creation of digital code to animate instruments, or in its anticipation in the theremin and its animation via hand-waving – is just one of many real and potential digital arts. The object of the art can just as well be painting, in which code can be written for colors, or colors can be effortlessly mixed on the screen – or words. The principles are the same.

I have already discussed at length in *The Soft Edge* (1997b, Chapter 11) the benefits to the author of freeing the word from its fixation on paper. Rather than negotiating with the typewritten page about whether this or that revision is worthy of retyping, the writer on the word processor can make any revision or correction that comes to mind. Just as with electronic music, the increase in sophistication of the instrument – the word processor in contrast to the typewriter – provides more prerogative to the mind.

This removal or reduction of physical obstacles or challenges to creation – whether paper or ivory keyboard or colors in oil or water – speaks to the issue of gatekeeping we considered in Chapter 10. Before there was any form of political, social, or economic gatekeeping, there was and still is a gate far more profound and primordial, a gate of sheer physicality that presents itself to the mind every time it thinks, feels, imagines. Talk is cheap precisely because it is the one form of communication we come already equipped with. It costs nothing because it presents no physical gate to negotiate.

But the word "cheap" may belie a great advantage. Why is it better for the world that our dreams, our artistic accomplishments, remain private, hostage to a dexterity we may not possess with a physical instrument?

The eternal answer in support of the gatekeeper – whether of social or physical origin – is that such dexterities filter out creations that may be unworthy. Or, put otherwise, the argument on behalf of performance as gatekeeping is that the shifting of emphasis onto pre-recorded tapes or computers that can turn our thoughts into art may result in something mechanical, artificial, nonhuman.

I would argue just the reverse – because the source of the art is not only still human, but even more so in the sense that thinking, feeling, imagining seem more quintessential to our existence as human beings than running one's fingers over a string or blowing into a reed (by which I mean to take nothing away from the extraordinary human talent needed to accomplish such stroking and blowing – only that the thought behind it seems more primary and indispensable to our specieshood).

But what about the monkey on the typewriter scenario – the possibility that a person with no mental talent can use sophisticated digital instrumentation to accidentally produce art?

Aside from, "so what?" – people are always free to ignore a poor production – I do not think this is real possibility. Taking myself as an example, I have known ever since kindergarten that I have no talent in drawing and painting. The most sophisticated virtual drawing program in the world would be useless for me, because I have not the interior mental talent to make it work. At the same time, I have arranged and performed vocal harmonies in rock music, written and produced songs, and tinkered around with various keyboard instruments since the 1960s. My guess would be that, given the time, I might produce something better than passable on a MIDI or other digitally-connected instrument. The difference between the drawing and the music is that I already have the music in my head, and would therefore be in a position to benefit from the removal of whatever physical barriers present themselves to the music's production.

Of course, ease of production, even with the thought behind it, is not without risks. In the virtual worlds of business and education I have plied in the past decade and a half, I sometimes find that my capacity to generate projects and e-mail outstrips the possibilities of their real fruition. I can write and send e-mail faster than its recipients may be able to act upon it. The physical barrier to my mind's eye in that case – another human being – is not amenable to bypass by a more sophisticated machine, at least not in any immediate future.

This false sense of accomplishment which the virtual world can engender when there is, in effect, no printer to render my word-processed text into words on paper (or when there are no computer screens on which other people can read my words), no converter of digital code into music, or waves of the hands into sound, is certainly a problem that we will need to address in the new century. Media evolution is notoriously uneven. In a world in which digital facilitation has increased the number of tasks not only that we can do but imagine that we can do, it is almost a foregone conclusion that the equivalent of printers and sound speakers in some parts of some systems will not be available. One reason why the Balinese might have been able to do everything well, after all, is that there were not that many things to do in their pre-industrial society. Further, the need to master fewer implements meant that the physical rendition of mental constructs was easier – or at least less problematic, since some mental constructs, like flying to the moon, were ipso facto recognized as impossible, and thus limited to fantasy, as of course they are not in our society.

But the main difficulty in our becoming virtual virtuosos, even when the result can be implemented in reality – i.e., when there is equipment at hand sufficient to convert our thoughts into print, music, pleasing and useful things outside of our brain, and people in a position to respond to these communications – is that, however much this may accommodate our imaginations, this is to a large degree at odds with the way we have been doing things for a very long time. Word processing and its products are easier for the world to accept than electronic music because the writing process prior to computers was in its external aspects not that different – to the non-writer, pounding away on the keyboard of a typewriter seems much the same as pecking on the keys of a computer. But typing on a computer, or waving one's hands before loop and rod antennae, seem very different from picking a guitar or whistling into a flute. The computer, in these cases of music, collides with our seemingly innate sense of how music should be made.

That sense is obviously not innate but learned, from centuries, even millennia, of tradition. But it is no less powerful. We have seen it glinting, lurking behind many of the digital media developments we have been considering in this book. Its spell is our constant companion as we think about and perhaps come to use new inventions – it exerts its pull on every media transaction.

In the next chapter, we give our full attention at last to McLuhan's "rear-view mirror."

14

THROUGH A GLASS, BRIGHTLY

Stroll into any store that sells new musical instruments these days, and you'll see an assortment of electronic instruments including pianos, organs, guitars, and saxophones. Except, upon closer inspection, you'll find that some of these electronic pianos are in no sense pianos – they have keyboards all right, but the keyboards are not connected to strings or anything, amplified or otherwise, that directly makes music. And the same is true of some of these guitars and saxophones – you can pluck and strum and blow into them, but that will not really be how their music is produced. Rather, the music that comes out of them – either then or eventually – will result from digital code that activates other sounds. For these pseudo guitars and virtual saxophones are but digital input devices; and the keyboard of the digital piano as well is much closer in function to the keyboard of a computer than the keyboard of a traditional piano. All of these are ways of entering digital data that will be transformed into music.

But why should they look like pianos, guitars, and saxophones? Surely the digital data they generate could be written on a personal computer.

McLuhan provides an answer: "We look at the present through a rear-view mirror. We march backwards into the future" (McLuhan & Fiore, 1967, n.p.). "Faced with a totally new situation," he explains, "we tend always to attach ourselves to the objects...of the most recent past."

The situation need not be totally new, and the attachment is linguistic as well as physical. In the rear-view mirror band that we encounter in our music store, we not only see instruments that are made to look like pianos, guitars, and saxophones, but we call them that as well. And even though their names have a digital prefix or a MIDI suffix, which changes their very mode of music production, we are nonetheless more inclined to think of them as just souped-up versions of what we already expect of musical instruments.

As is the case with almost all of McLuhan's ingenious labels for profound relationships in media, once you begin looking for rear-view mirrorisms, you see them everywhere you turn in history. The telephone was first called the talking telegraph; the automobile the horseless carriage; the radio the wireless. In each of these cases, the proximate effect of the rear-view mirror was to obscure some of the most important revolutionary functions of the new medium. Thus, although the telephone of course indeed talked, it was also situated inside the home, a privatization and personalization of the telegraph that transformed family and business – not even hinted at in the label "talking," which can occur just as easily outside as inside of the home. Although the automobile was horseless, this negative appellation said nothing specific about the combustible engine that would go on to make oil among the most precious and contested commodities in the twentieth century. Nor did the name "wireless" – although radio indeed had none – suggest in the slightest the simultaneous mass audience which radio would bring into being. For that audience was a consequence not of radio's lack of wires, but of the prohibitive expense of putting radio transmitters into everyone's home (in contrast to the much less expensive transmission component of the wired telephone).

Indeed, the rear-view mirror becomes, for McLuhan, a fundamental operating principle for the evolution of media and its effects, an alternate snapshot that encompasses and overlaps with many of McLuhan's other essential insights that we have already considered in this book (again, not surprising, in view of the holographic nature of McLuhan's ideas, wherein each one provides a blueprint and point of entry for the others – an ideational hypertext link, ahead of its time). Thus, McLuhan's notion of older media becoming content for newer media, and therein becoming more visible to the point of being mistaken for the newer media, is but a rendition of the rear-view mirror, and its redirection of our gaze ahead to the just-passed. So too is his observation that we fall in love with ourselves when we look into our reflections in our media, and become blind to their real effects, for we are the media's proximate creators. Donald Theall –

though he and McLuhan disagreed on many issues – nonetheless caught this centrality of the rear-view mirror early on, entitling his critical analysis of McLuhan's work, *The Medium is the Rear-View Mirror: Understanding McLuhan* (1971).

So pervasive is the rear-view mirror as a mode of thinking about the present and future, that McLuhan himself was prone to it, as he was the first to admit. Thus, his notion of the global village is itself of course a rear-view mirror, or an attempt to understand the new world of electronic media via reference to the older world of villages (although the village is older than the city, and thus an example of a deep rather than "recent" past, which McLuhan sees as the prime rear-view mirror territory; we will discuss McLuhan's view of such deep retrievals, and their differences from the rear-view mirror, in the next chapter). Indeed, McLuhan's reliance of metaphor made the rear-view inescapable as a driving mechanism of his work. As we saw in Chapter 2, he was fond of quoting the anonymous "wag" who punned on Browning that "Man's reach must exceed his grasp, or what's a metaphor?" (e.g., 1977a, p. 176; see also McLuhan & Parker, 1968, p. 12). But, indeed, what *is* a metaphor, if not a rear-view mirror, that seeks to make the new, the unclear, the less known more explicable by attaching it to something we feel we already know, inside and out?

But whereas McLuhan thought he could control the possibility for misunderstanding also inherent in metaphor – even if many of his readers could not – he saw the rear-view mirror as so irresistible and prevalent in our treatment of media and technology as to be unavoidable. The best he thought he and we could do was identify its presence wherever possible, and indeed he often cited his position in Canada, above the "DEW line" (Distant Early Warning system) of cutting-edge media in the United States, as giving him an advantage in those identifications (e.g., McLuhan & Powers, 1989, Chapter 10). In that spirit, we begin our scrutiny of the rear-view mirror behind our computer screen, and its distortion of our perception and understanding of the digital age.

THE WEB OF REAR-VIEW MIRRORS

The Web is a veritable hall of rear-view mirrors, because we do so many new and old things upon it, in new ways. When we listen to RealAudio on the Internet, how can we not think of it as radio? When we do research upon it, is it not a library? Where is the distortion in treating an online chat room as if it were a cafe?

In an important sense, these rear-view mirrors are *not* distortions, and their examples call attention to an important benefit of walking into the future with our eyes upon the past: the rear-view mirror, like metaphor, helps us come to terms and feel comfortable with new media. We all know the value of books. To point out that reading text online is in many ways the same as reading it in a book, to call the Internet an online book, is to underline its value to the world at large, including critics whose focus on other aspects of the Internet may lose sight of its benefits.

But the rear-view mirror is a two-edged sword, and in that sense the critics are not wrong in some of their concerns. Parsing metaphors or rear-view mirrors is notoriously difficult, and often the mirror may blind us to ways in which the new medium is not analogous to media of the past. When we use the Web as a library, and our connection freezes or Windows crashes, we may be literally unable to continue reading the text; that simply cannot happen with a physical book in hand, unless we're struck sightless. If we come to know someone online, even see photographs or live video images of his or her face, we still do not know with certainty various aspects of that person which we could perceive in an instant in an in-person encounter: the photo or even video image, for example, could be a phoney (an instructive word, by the way, because its derivation apparently lies in the possibilities of deceit in a telephone conversation). We might well enjoy RealAudio on a palmtop computer as we recline on a beach, waves lapping near our feet, much as we might enjoy an equivalent broadcast from a little battery-operated radio that cost us under ten dollars. But the palmtop today costs ten to a hundred times more, and a big wave washing away the palmtop would be excruciating for most of us, financially, in comparison to our losses were a similar wave to hit the radio.

Our problem, then – if I may put the rear-view mirror back into its original environment in the car – is to know when to stop looking into the rear-view mirror, and when to look at the road ahead. If we stare too long into the rear-view mirror, focusing only on how the new medium relates to media of the immediate past, we may crash head-on into an unseen, unexpected consequence. On the other hand, if we look only straight and stiffly ahead, with no image or idea of where we're coming from, where we've just been, we cannot possibly have a clear comprehension of where we are going.

Our situation is further complicated by the fact that, if we use the rear-view mirror to help navigate our vehicle into the future, we not only need accurate information of the past – a reflection of where we have just been

– but some indication of, or way of judging, which aspects of that past are most relevant to our journey ahead. A stationary maple in the rear-view mirror is usually much less relevant than a Corvette overtaking us at 100 miles-per-hour on our right.

Speaking of stationary and stationery, consider the case of "electronic ink," currently under development at M.I.T. (see Peterson, 1998, for details). The idea is that we will be able to write on sheaves as thin and flexible as paper. But the "paper" will be coated with electronically charged particles, whose status can be easily altered by a specific electronic field, thus allowing the writer with the proper equipment to erase or revise – and indeed enjoy all the benefits of text liberated from its fixation on traditional paper pages (see Levinson, 1989, for some early thoughts on the electronic liberation of text).

A quick glance in the rear-view mirror might suggest that such electronic ink is an ideal solution: it allows the convenience of paper, with the word processing and telecommunication possibilities of text on computers with screens. But, on more careful examination, we find that we may have been looking at not the most relevant part of the immediately past environment. For one of the great advantages of words fixed on traditional paper is indeed that they are stationary, with an "a": we have come to assume, and indeed much of our society has come to rest upon the assumption, that the words in books, magazines, and newspapers will be there for us, in exactly the way we first saw them, any time we look at them again in the future. Thus, the stationery as stationary, the book as reliable locus, is a function at least as important as their convenience in comparison to text on computers (see Levinson, 1998a, for more). Of course, we may in the future develop electronic modes of text that provide security and continuity of text equivalent to that on paper – modes that in effect allow the liberation of text without any diminution of its reliability – but current electronic "inks" and "papers" are ink and paper only via vision in a rear-view mirror that occludes a crucial, desirable component of the original.

Invocation of desirability brings to the fore a facet of media analysis inherent in much of this book – but, as we have seen, explicitly denied as a goal of media studies by McLuhan (e.g., "value is irrelevant," in Stearn, 1967, p. 286), who claimed to eschew not only value-judgements but explanation in his explorations of media (see also Chapter 2 in the current book). As I have argued here and elsewhere (e.g., Levinson, 1981a), however, McLuhan not only explained but did offer evaluation throughout his explorations: when he observes that "the increase in visual stress among the Greeks alienated them from the primitive art that the new electronic age now reinvents" (1962, p. 81), he is surely saying something evaluative, more specifically, negative, about what he thinks happened to the Greeks. And this dimension of evaluation is all to the good in McLuhan's work, even if it was denied by him.

Thus, in the spirit of what McLuhan actually did rather than what he said he did, we in the next section escort his thinking about rear-view mirrors into the new millennium by discussing how they can be used to better alert us to benefits and drawbacks of new media.

REAR-VIEW MIRROR IN THE TOOLBOX

The key to human improvement of media is control, to wit, our control of them. Although we can and do and did and will make mistakes – many of them – I take it as a given that a technology in our control is preferable to a technology beyond our power to choose to use in this way or that. For indeed, to the degree that technologies are beyond our control, we are no better than the least intelligent organism in its relation to the environment. If someone wishes to contend that such organisms are better in the grand scheme of things, even better for the Earth, than we humans, and that therefore our increased levels of control are not an improvement but a degradation, I will happily concede to that person that there is no point in continuing this discussion. But if you believe as self-evident that human rationality is in some profound sense an improvement over its predecessors on Earth, then read on.

Whatever McLuhan believed about rationality – he described it as a consequence of "the coded, lineal ways" of the Greeks (in Stearn, 1967, p. 270), i.e., of their alphabet, and was quite rational in his bringing to bear of evidence in support of his propositions (see Levinson, 1981a) – he was prone to view the whole and the particulars of human history as the consequence of technological effects over which we have little control and often no understanding. Indeed, his emphasis of such lack of control makes his protestation of values more explicable: if we are truly numb to the most profound effects and impact of media, if we are no better than Narcissus who died in the hypnotic pursuit of his own reflection (McLuhan, 1964, Chapter 4), then what is the point of attempting to determine which medium is better or worse for us? We are powerless to do anything about such determinations anyway.

But the history of media, and our development in particular of what I call "remedial media," demonstrates that we indeed have some control. We invented windows (with a small "w") as an improvement upon the poor choice of walls with no vision or walls with holes and no warmth; when we later discovered that this new medium had the unintended consequence of empowering the Peeping Tom, we improved the invention again with window shades. We invented Windows (with a large "W") as improvement upon the choice of DOS's boring precision and Macintosh's playful fuzziness; as we discover problems with Windows, we will likely invent the equivalent of Windowshades (which is why, to harken back to Chapter 7, anti-monopoly suits against Microsoft are so irrelevant: its products will succeed only so long as they are useful). And windows and Windows are just two of numerous examples of our rationality devising improvements in response to problems in media. We disliked losing phone calls when we were not at home to receive them? We invented the remedial medium of the telephone answering machine (actually, first envisioned by Edison as the primary use of his phonograph in 1877, a year after Bell filed his patent for the telephone). We regretted being dictated to by the schedule of television? We invented the VCR to liberate us from this regimen (see Levinson, 1997b, for more on specific remedial media; see Wachtel, 1977/1978, for early discussion of the window as the archetypal medium).

Indeed, as I discuss in my "anthropotropic" theory of media evolution – tropic=towards, anthropo=human – the overall evolution of media can be seen as an attempt, first, to fulfill the yearnings of imagination by inventing media that extend communication beyond the biological boundaries of hearing and seeing (thus, hieroglyphics and the alphabet and the telegraph each in its way extends words thousands of years and/or thousands of miles), and, second, to recapture elements of the natural world lost in the initial extension (thus, photography recaptures the literal image lost in writing, and telephone, the phonograph, and radio recapture the voice). From this vantage point, the entire evolution of media can be seen as remedial. And the Internet, with its improvement of newspapers, books, radio, television, et al. can be seen as the remedial medium of remedial media (see Levinson, 1979a, for the first sustained presentation of "anthropotropic" media; I first discussed remedial media in Levinson, 1988b, pp. 225–6).

Thus, I see the phenomenon of remedial media, and the prevalence of human rationality and control manifest in its application, as fundamental in the evolution of media and their effects, as much of a major factor as the unintended consequences of media – the effects of technology to which we are blind – which McLuhan devoted most of his professional life to studying and elucidating. In McLuhan's parlance, the evolution of media is thus itself cool – often below the surface, low in profile, but always inviting our ministrations for improvement.

With the possibility of such media-steering at hand, I think it eminently logical to look upon the rear-view mirror not just as a post-mortem of a mistaken perception, not just a shorthand for our inability to see the present and the future as it really is, but also as a device to help us actively guide our technology to better paths.

How, more specifically, might we consult the rear-view in this way?

Neil Postman (e.g., 1998) often says that we ought to ask ourselves, of a new technology, what problems does it solve or seek to address for us? He intends the answer, upon careful scrutiny, to be few or none of any importance, with the consequent conclusion that we embrace our technologies for false or frivolous reasons, leading to at least a waste of time and at most a threat to our well-being. I propose the rear-view mirror as a tool for coming up with better answers to Postman's question.

The rear-view mirror, as a projector of our immediate past merging into the present, is an ideal device for keeping us abreast of our real problems. In clear-cut cases of remedial media, such as the VCR, we have no need of a rear-view mirror, because the problem at hand and its remedy are so obvious: everyone was aware of television's drawbacks as a provider of evanescent, unprogrammable content; the VCR was invented to remedy those shortcomings; and everyone immediately recognized the VCR as such. But media, as we have seen, also traffic extensively in unintended consequences, which means that the problems solved in those cases may not at all be immediately apparent. We may, indeed, have a medium at hand, and be utterly unaware of the profound problems it addresses. Thus, Bell invented the telephone largely in pursuit of a hearing aid for his wife; Edison (as noted above) at first thought his phonograph would work primarily as a telephone answering machine; and a decade later, after he had come to realize that the phonograph solved the problem of providing permanent musical entertainment, Edison initially envisioned his motion picture process mostly as a visual adjunct to the phonograph (see Levinson, 1997b, for details). In such cases – which seem to far outnumber direct cause-and-effect remedial media like the VCR – the rear-view mirror becomes an essential tool for telling us which initial perceptions to favor.

The Internet provides a sterling example of a medium brought into being for reasons that have little to do with the variety of profound problems it addresses. Indeed, its origins as the military online network ARPANET are diametrically opposite its smashing of gatekeeping and hierarchies which we have explored throughout this book. Created to facilitate communication within a highly developed hierarchy, the Internet went on to increase communication to such an extent that it levels hierarchies. This was unapparent to its initial funders. But theorists such as Vannevar Bush, Douglas Englebart, and Theodor Nelson – in touch with profound problems in communication which needed remediation, in effect using a rear-view mirror to keep in contact with these problems as they explored the possibilities of new media – were to some degree able to foresee the real benefits of the Internet (see Skagestad, 1993, 1996, for more on Bush, Englebart, and Nelson). Among the problems that the Internet has alleviated, in addition to gatekeeping, are the slowness of communication on paper, the shutting out of many minds from what Comenius called the Great Dialogue when it is conducted offline, the discrepancy between good ideas in the mind and the difficulty of giving them expression in tangible reality, and many other shortcomings of prior media discussed in this book. All become clearer in the rear-view mirror.

Which is not to say that all problems become solvable, or even crystal clear, in the rear-view mirror. It is not Prester John's Speculum, capable of detecting every falsehood, distortion, and conspiracy in the kingdom of media. A glint of sunlight bouncing off of a less relevant component of the immediate past can momentarily blind us to a more relevant component, as we saw with the "wireless"; or, we may coolly scrutinize a past environment, and still overlook a crucial feature, as we saw with "electronic ink" and its erasure of print's reliable inability to be erased. In other words, like every technology and every theory, the rear-view mirror ought not be used uncritically.

But if we take the rear-view mirror with the visual equivalent of a grain of salt – "caution: objects in mirror may be closer than they appear" (my son Simon quoted this common warning on the passenger's-side mirror to me, when we were discussing the need for caution when applying McLuhan's rear-view mirror) – we have a most helpful, even necessary, device for giving us a measure of the future. For the future can be rationally measured only in terms of the past (the present, on this reading – or driving – being a hypothetical midpoint between immediate past and immediate future).

McLuhan, for the most part, was concerned more with the past leading up to the present than the present unrolling into the future. He was first and foremost an historian, and his most vivid metaphors either utilize a past environment for instructive comparison (e.g., global village) or are borrowed from one (e.g., hot and cool from the jazz age). His critics usually miss the depth and encyclopedic scope and detail of his historical reach, or they twist this virtue into a liability, as in Dwight MacDonald's infamous "he has looted all culture, from cave painting to *Mad* magazine, for fragments to shore up against the ruin of his system" (in Stearn, 1967, p. 203). But McLuhan's critics have not been entirely wrong in their frustration that his work, in the form he presented it, does not provide a clear guide to the future. Because he was not trying to be a guide. To explore and to guide are two quite different activities.

But to bring McLuhan's work into the future, as this book is attempting to do, is inevitably to shift at least some of its focus from the past to the future. For, at very least, the application of McLuhan's constructs to current media and their effects is to direct those constructions, created in the 1950s, 60s, and 70s, to the beginning of a new century and millennium – that is, to utilize those constructs in the exploration and explication of their future.

In the next section, we examine some of the hidden prospects and difficulties of marshalling not only the rear-view mirror but other of McLuhan's insights for help in charting our future. (OK, I admit it; it's the worst possible pun in these circumstances. But you knew I wouldn't be able to resist it sooner or later. And I did, after all, manage to control myself until this, the next-to-last chapter in this book.)

PRESCRIPTIONS FOR THE FUTURE

I returned home from vacation in August 1978 to find the most delightful – and instructive – message from Marshall McLuhan on my telephone answering machine. "I am enjoying the dissertation," he said, knowing I would instantly recognize his voice without identification, which of course I did, "but you misrepresent Innis and me when you say that we are 'media determinists'… ."

The dissertation in question was mine – "Human Replay: A Theory of the Evolution of Media" (Ph.D. diss., New York University, 1979a) – and I also recognized the substance of his message as a courteous variant of the species of responses by McLuhan to discussions of his work seen most publicly in his response to the pompous fool of a professor in *Annie Hall* (1973), who was insisting that television was a hot medium. Mixed with my genuine joy about getting any response at all, let alone a partially positive one, to my doctoral dissertation from McLuhan – who by this time was not only someone I regarded as by far the most important thinker about media in the twentieth century (and I still do) but a dear friend – was a firm conviction that I nonetheless was correct in my appellation of Innis and McLuhan as media determinists, and McLuhan was the one who was mistaken in this instance (modesty was never one of my virtues – if it is really anyone's).

The section of my dissertation which had attracted McLuhan's attention – I had mailed the manuscript to him just a few weeks earlier, at the same time that I submitted it to my doctoral committee at New York University, upon which Neil Postman served as my primary thesis adviser and Chair – was my description as "media determinist" McLuhan's view that "man becomes, as it were, the sex organs of the machine world...enabling it to fecundate and to evolve ever new forms" (McLuhan, 1964, p. 56; see also discussion of this observation in Chapter 3 of the current book). Significantly, the difference between McLuhan and me on this issue was more of emphasis than substance. My dissertation was, after all, a theory of the evolution of media – that theory being that human beings select for survival the media most appropriate to our needs. So I indeed agreed with McLuhan that we enable media to fecundate, and I in fact developed that insight into my "anthropotropic" theory (see earlier in this chapter). But McLuhan's rendition of this insight, coupled with his view that we hypnotize ourselves via our media ("Narcissus Narcosis," 1964, Chapter 4), indicated to me that McLuhan's take on the human relationship with media was that we were its products or effects, rather than vice versa. Hence, my description of this view (and other similar views of Innis and McLuhan) as "media determinist": although humans clearly have some control over media, the two saw media as calling the shots. New media turn older media into art forms; new media recreate the world as a global village; current media become rear-view mirrors that blind us to the impact of new media; and while all of this and more is happening, we focus numbly and dumbly on the content. In other words, according to McLuhan, the characteristic of media that counts most is that they regulate information and determine events while we stare at inconsequential, attractive displays and pamper ourselves with the illusion that we are in charge.

In 1978, this seemed to me ipso facto evidence that McLuhan was a media determinist. Now, with the wisdom of hindsight – in effect, looking back at McLuhan and my initial work about him in my own rear-view mirror – I can see that, although I still think those views put humans into an inferior position vis-à-vis technology, the choice of "media determinism" may not have been the best label to describe them. I still disagree with McLuhan on the emphasis on technology, and diminishment of human control, entailed in those views. But I understand why he resisted the appellation of media determinist. That resistance had to do with his abstention from predicting the future.

As Karl Popper demonstrated in detail in his critiques of Marxism (e.g., Popper, 1945, 1957), there is a significant difference between the study of history, which attempts to describe and explain the past, and "historicism," which attempts to derive from such explanations an overarching theory of human societies and their evolution. The latter approach, amply exampled in Marxism, easily moves from identifying a central human activity that shapes history to claiming an inevitable future of some very specific sort based on the continued operation of this activity in the expected way. Thus, Marxism claims that economic relationships determine all others in society; posits that in the most advanced societies, workers will come to grasp that truth and take control of their lives and societies; and contends that inevitably the revolution will spread to the whole world. But this, of course, was not how it came to be. Already in the 1950s, even in that high-tide of Communist/Marxist political power in the world, Marxist predictions of how the future would arrive were unfolding backwards: Russia and China were among the least advanced, not most advanced, in economic development when their revolutions occurred. Lenin earlier had explained this deviation from Marxist prediction in the Russian experience as due to the unexpected benefits to the West of its imperialism, temporarily invigorating those advanced capitalist societies in a manner unforeseen by Marx. With the fall of the Soviet Union at the beginning of the last decade of the twentieth century, however, it became clear that Marxism and the overthrow or withering of capitalism would be the wave of the future for nobody. Other than deficits specific to Marxist theory, the lesson – offered by Popper in the 1940s and 50s, as a philosopher of science – is that the future is essentially unpredictable by a monolithic determinist theory. There are simply too many variables in human life. (Interestingly, Isaac Asimov came to the same conclusion in his science fiction classic, the *Foundation* series, in which a field of "psychohistory," which attempts to chart and control the future based on statistics that predict mass behavior, ultimately fails, or works only with active interventions which take account of events unforeseen in the psychohistorical projections; see Asimov, 1951, 1952, 1953.)

Leaving aside the question of to what extent Marx was a Marxist, we can see that McLuhan was a media determinist – substituting media, or the way that information is managed, for economics, or the way that labor and wealth are managed – only in the first part of the Marxist schema. That is, McLuhan discerned and explored the ways that media determine all other aspects of society – including politics, art, education, business, and the many other activities considered in this book. But McLuhan derived no specific predictions whatsoever about inevitable futures. Indeed, he offered no overarching theories about *how* media operated in the past – no single direction in which he saw their impact – except to emphasize that their operation was and is extraordinarily important, with the result of frequently overwhelming our human powers of choice. Thus, rather than a theory, and the predictions that theories ineluctably imply, he left us insights – the medium is the message, the rear-view mirror, etc. Insights of course can generate given predictions, but unless the insights are all part of some grand theory, there is no reason to assume that such predictions will even point in the same direction.

I disagree with McLuhan in this area in two ways. And yet, the disagreements lead to the same conclusion about an open, unpredictable, imprescriptible future – in effect, the two disagreements cancel each other out. First, I do see an overarching pattern to the evolution of media – to wit, that media evolve towards increasing consonance with pre-technological human communication modes, while maintaining their extension across time and space that our imaginations inspire (my "anthropotropic theory" of media evolution). But, second, since I see humans as fundamentally in charge of this process – sometimes with explicit, conscious application of rationality, as in remedial media – I do not see media, as part of this overarching pattern or anything else, as overwhelming our powers of choice and selection. We may be numb or mesmerized by media, but the effects are temporary. The thrust of most of this chapter has been to demonstrate how we can and have looked away from the rear-view mirror. David Sarnoff realized as early as 1915 that radio would work best as a mass-medium music reception box, and Theodor Nelson and others were quick to see the genuinely new uses of computers for communication. Further, certainly we can squint and shift our gaze within the rear-view mirror, as when we consider that paper has the advantage of being both convenient and permanent, and conclude that both need to be addressed when we consider the advent of electronic ink, indeed electronic text in general, and its future prospects.

Had I written this book when I first applied to the Ph.D. program at New York University in 1975, I could have ended it right here, with this chapter. But, unsurprisingly, McLuhan had one last surprise in store. It was not quite a theory of media. It was still – like hot and cool, light-through/light-on, the rear-view mirror – a diagnostic tool. But it was a tool unlike all the others. For it was comprehensive – not only about history, but about all of McLuhan's insights into media and history, which this tool attempted to encompass. And built right into this tool of tools was a mechanism for predicting the future. Not a single, grand, unified future – not at all a species of determinism that sees a definite way that the world must be – but a multiplicity, even a myriad, of futures from a myriad of technological possibilities, a kaleidoscope of futures in the multiple potentials of the media currently before us.

This tool, in other words, was surprising both in its attempt to reduce all of history to a series of common denominators, and in its pointing of these towards the future. But it was otherwise just what you would expect of McLuhan and his incandescent openness to possibilities.

The tool was first unveiled as such in a small piece published in *Technology and Culture* in January 1975. But I had not seen it as yet when I entered Neil Postman's office about two years later. Neil, in addition to being the guiding force of that doctoral program (and my dissertation adviser) was at the time also Editor of *et cetera*, a journal published by the International Society for General Semantics. Neil's brow was creased in the way that it often is when he is thinking hard about something; and a cigarette dangled from his fingers, less than half an inch from burning them.

"Paul." He looked up at me, and gestured me to a seat. "What do you think of this?"

He slid a slim manuscript across his desk to me.

It said: "Laws of the Media by Marshall McLuhan" – the subject of our next, and concluding, chapter.

15

SPIRALS OF MEDIA EVOLUTION

"When I came across Karl Popper's principle that a scientific hypothesis is one that is capable of falsification," McLuhan began his essay published in *et cetera* in June 1977, "I decided to hypothesize the 'Laws of the Media'" (1977a).

Here we see McLuhan's sense of humor and his continuing distrust of grand theories of human behavior that purport scientific significance – such as Marxism – both at their height: He allows himself to offer a theory, a set of principles or laws, that could be mistaken to purport such scientific significance, because, after all, such laws can always be shown to be false, and therefore not do too much damage.

So why present such a theory – or tool at least partially in the form of scientific laws – in the first place?

The answer, of course, is that McLuhan thought it had something of value to provide to our understanding of media. And, in fact, it did and does nothing less than tie together, even clarify, just about all of McLuhan's major insights – the amplification of acoustic space by electronic media, the obsolescence of print by those same mass (from our perspective, pre-computer) media, the retrieval of elements of the village on a global scale by those media, and their eventual reversal – something McLuhan did not live to see – into a very different kind of electronic milieu, the digital online age and its capacity for interaction and reduction of gatekeeping that we have today.

McLuhan's four laws or effects of media – amplification, obsolescence, retrieval, reversal – indeed became McLuhan's swan song. But they were only somewhat more clear than the disparate, interrelated insights like the global village and the rear-view mirror they brought together. And in many respects, their very open-endedness – in which a given medium, such as television, could reverse not only into computers but into cable, VCRs, holography, and almost as many things as a mind could reasonably imagine – made them just as frustrating as the earlier formulations for people with no heart for McLuhan's daring metaphors.

Neil Postman had been an admirer of McLuhan's style and originality since the 1950s. As I sat across that table in Neil's office in February 1977, reading the "Laws of the Media" manuscript, he again asked me, "What do you think of this?"

I explained what I was deriving from that article.

"Good," Neil said. "Why don't you write up a brief description of what you just told me, and I can pass that on to whoever edits this article for *et cetera*."

In fact, Neil went on to append my description to the front of McLuhan's article, with the intention of publishing it as a Preface. He sent this Preface up to McLuhan, and much to my amazement and delight, McLuhan was not unpleased with it. The article with Preface was published in June 1977 – we had met for the first time about a month before that – and in March of the next year Marshall and his son Eric came to Fairleigh Dickinson University, where I had organized a conference on the "tetrad," McLuhan's favorite alternate title for his four laws of media.

But the import and meaning and intended mode for applying the laws were still far from clear.

McLuhan was hard at work on a book-length manuscript about the laws when he suffered a major stroke in September 1979. Doubleday, the putative publisher, had been expressing unhappiness with the manuscript since its agreement to publish McLuhan's book on the laws of the media in 1974. Marchand (1989, p. 243) quotes Betty Corson, Doubleday's editor-in-chief in Canada, as saying the book "was not in publishable shape at the time I saw it"; but I saw the manuscript in the late 1970s, and it was no less readable than any of McLuhan's other books – meaning, yes, it was written in McLuhan's inimitably aphoristic style, but it was a treasure chest of insights for those willing to take the time. In any case, McLuhan's death on New Year's Eve, December 31, 1980, effectively ended Doubleday's responsibility to publish the book.

I called Doubleday's editor for the project in New York, Loretta Barrett, the first week in January, and urged that they go ahead and publish the book. I followed up with a detailed two-page letter on January 7, in which I said "that publication of the *Laws of Media* by Doubleday would be a most significant event in media scholarship, and likely to be very eagerly received" (Levinson, 1981e). Doubleday was unpersuaded. McLuhan's reputation, already in eclipse at that time, and my reputation, negligible at that time in media scholarship, were not enough to open the gates at Doubleday on that occasion.

It remained for Eric McLuhan to take the manuscript in hand, and see it through publication by the University of Toronto Press in 1988 (M. McLuhan & E. McLuhan, 1988). I reviewed it – along with Marchand's biography (1989), McLuhan's *Letters* (Molinaro, C. McLuhan & Toye, eds., 1987) and *The Global Village* (McLuhan & Powers, 1989) – for the prestigious *Journal of Communication* in 1990 (Levinson, 1990). Four years later, I published a streamlined version of the review in the third issue of *WIRED* (1993), which was listing McLuhan as its "Patron Saint" on the masthead.

By then, the digital revolution had retrieved McLuhan's work and reputation from its temporary burial by obsolesced mass media, as we will see in detail below...

BASICS OF THE TETRAD

McLuhan's tetrad asks four questions about the impact and development of any medium: What aspect of society or human life does it enhance or amplify? What aspect, in favor or high prominence before the arrival of the medium in question, does it eclipse or obsolesce? What does the medium retrieve or pull back into center stage from the shadows of obsolescence? And what does the medium reverse or flip into when it has run its course or been developed to its fullest potential?

The operation of these four laws or effects becomes clearer when we consider the circumstances surrounding any medium. Radio, for example, amplifies the human voice instantly across vast distances to a mass audience. It obsolesces print as a mass medium, as when, for example, we receive our first word of an important news event on radio rather than via an "extra" addition of a newspaper. It retrieves the town crier, who had been obsolesced to a large extent by print. And acoustic radio, when pushed to its limits, reverses into audio-visual television.

And the same process commences with the next medium – television –

the medium that radio reversed into. TV amplifies the visual, but in an "acoustic" all-at-once sense, not in the one-on-one sense of individuals reading separate newspapers, likely not all on the same page. TV obsolesces radio, obviously. It retrieves the visual – which also may seem obvious by now – but not in the way the visuality of print had been obsolesced by radio. The retrieval of the visual in TV is rather something new, a hybrid of previous visuality with current electronic attributes that is genuinely different. And when limned to its fullest extent, the screen of television flips into the screen of the personal computer.

The more we consider even just these two examples, the more we see things that may not be so obvious. TV, for example, obsolesces not only radio but motion picture theaters: the audio-visual in the home not only replaced radio as a narrative audio medium, but resulted in a sharp drop in the number of movie theaters, and the number of films thus available to the public in such theaters. Radio, for its part, obsolesced not only visual print, but some aspects of non-electronic conversation: to hear a voice on the radio is not the same as speaking with someone in person. And, as already indicated above, television flips into many more media than just the personal computer. Two-dimensional television reverses into three-dimensional holography. Evanescent television reverses into the VCR: TV with a memory. The oligarchy of network TV (a carry-over from radio) and its severely limited number of channels reverses into the multiplicity of cable.

One general lesson here – one of several general lessons about applications of the tetrad and how to get the most benefit from them – is that the four effects of the tetrad are rarely singular. Instead, given media usually enhance, obsolesce, retrieve, and reverse into many things. Further, more than one medium may enhance, obsolesce, retrieve, or reverse into the same thing. Television reverses into the computer, but so too, in a different line of tetradic development which we have explored considerably in this volume, does the book. Indeed, so does the telephone. And, when the personal computer is used for live chats online, it can also be seen as the fourth result of a tetrad for CB-radio – what it, too, flipped into in the digital age.

The notion of a tetrad line of development brings to the fore another significant feature of this four-way mode of analysis – another important lesson in how to best use it. As I pointed out in my 1977 Preface to McLuhan's "Laws of the Media," there is a cyclical but progressive relationship among media and their effects which becomes plain when they are parsed according to the four "laws." What radio obsolesces – visuality –

television retrieves. And in so doing, television – what radio has flipped into – obsolesces the purely acoustic radio. There is a circularity of sorts here, which led me to term this ongoing movement of media "Tetradic Wheels of Evolution" (Levinson, 1978b) in the paper I delivered at the 1978 Tetrad Conference with Marshall and Eric (and Robert Blechman and Jim Morriss) at Fairleigh Dickinson University. But actually, as I explained in that paper, there is a real movement forward in this process – it is not just a circle – and so it might better be termed a spiral. What television retrieves, as I indicated above, is a genuinely original compound of prior environments with some wholly new properties. Or we might say that although the reversal of radio into television retrieves what was obsolesced by radio – in this specific example, the visual – that rescued environment runs differently when it is enhanced by the new medium (television) than it did before it was obsolesced.

Although these and other significant aspects of tetradic analysis were inherent in McLuhan's first (and only, in his lifetime) presentations of them (1975, 1977a), they were not then spelled out or explored. Indeed, even the Tetrad Conference at Fairleigh Dickinson University went largely unnoticed by the scholarly world, although James Curtis (1987, pp. 10–13) did a fine job of applying McLuhan's laws of media via my tetradic wheels of cultural evolution in Curtis' book on rock 'n' roll. The eventual publication of *Laws of Media* by Marshall and Eric (M. McLuhan & E. McLuhan, 1988) does have a brief discussion of "clusters" ("where a group of tetrads...reverse into the same mode of culture") and "chains" ("when...one tetrad's reversal...provides the enhancement...of the next tetrad") (p. 130; examples on pp. 208–14; the discussion and examples pertain to all four laws, not just reversal). These are the equivalent of my intersecting wheels or spirals – but much more remains to be done.

We join to the task in the next section, with special attention to the implications for media in the coming century.

SPIRIT OF THE DIGITAL AGE

It surely is no coincidence that multiplicity is the active ingredient in the cable TV that network television reverses into, the Web that books and libraries flip into, and indeed the computer screen that also succeeds television. Such an increase in choice across the board in our media apparently works like a basin of attraction – the more media it coaxes into flipping into this, the stronger it becomes.

Hegel's phrase for such convergences of effects was spirit of an age. And, indeed, in addition to thus recognizing what the McLuhans called clusters, Hegel devised a better-known tool for assessing human culture and activity in ways that bear considerable resemblance to the tetrad: the dialectic. Clearly, Hegel's famous tripartite parser covers much the same ground as the tetrad: the synthesis in effect retrieves the thesis which had been earlier obsolesced by the antithesis; and when the synthesis goes on to serve as the new thesis and generates a new antithesis, it is performing in a manner akin to what the tetrad calls reversal.

Here a significant difference does arise, since reversal, as we saw above, retrieves elements of the formerly obsolesced, and thus has some resonance with the past, even as it moves into the future. In contrast, the new antithesis of the dialectic need not have any connection at all to what came before the new thesis (or the previous synthesis): all it has to be is the opposite, in some profound way, of the new thesis. In that sense, the dialectic has a wild, revolutionary flair – no necessary nod to the past – even before Marx stands it on its head, whereas the tetrad insists on weaving originality with the already-known. This is consistent with both McLuhan's use of metaphor and the rear-view mirror, and makes the tetrad a much better mesh for sifting through history, and a more meaningful projector of the future, which becomes explicable as a replay of human patterns in new forms.

Of course, Hegel's work came first. Resemblances of the dialectic to the tetrad thus are due to at least some Hegelian thinking by McLuhan, despite his attempts to distance himself as much as possible from logic, philosophy, and indeed the tetrad from the dialectic. One of the many welcome notes he sent me on the subject in 1977 – imagine, here I was just finishing the first year of my doctoral studies, and I was getting almost weekly letters from someone Tom Wolfe had aptly compared to Darwin, Einstein, and Freud – took issue with me, in particular my article "Toy, Mirror, and Art" (Levinson, 1977b), for reliance on Hegel's three-part schema. "Just a note to suggest that your triads could become tetrads in all cases," McLuhan advised (8 September 1977), "i.e., each of toy, mirror, and art is a tetrad, and all together they form a triad which is 'missing' one term. On page 159 the retrieval factor (number 3) has been suppressed out of deference to Hegel?" (McLuhan, 1977b). (On that page I discussed the dialectic similarities of "Toy, Mirror, and Art" – the three developmental stages I saw in technology – to Jean Piaget's sensorimotor, concrete, and formal stages of intellectual development, McLuhan's oral, written, and electronic eras of communication, Freud's oral, anal, and genital stages of sexual expression, Walter Ong's comparison of Freud's and McLuhan's three stages, and Arthur Koestler's Jester, Sage, and Artist as the three unfolding stages of creativity; see Chapter 11 in this book for discussion of other facets of "Toy, Mirror, and Art.")

McLuhan was right that Hegel's dialectic gives short shrift to retrieval – with the result, as I indicated above, that it provides no deeply historical basis for projecting the future. But that is no reason to abandon the dialectic in favor of the tetrad, or even to prefer the tetrad in all cases. Triads are, after all, more concise than tetrads, and for that reason alone can be preferable as a short-hand. To say that the digital age is a synthesis of the individual printed book or newspaper and the mass-broadcast radio or television is meaningful and useful, and a search for the new antithesis provoked by this new digital thesis inevitably looks in the same direction as reversal, albeit without the tetrad's historical sweep or depth.

McLuhan came up to me with a smile after I had offered views like the above about triad and tetrad to his Monday night seminar in the Coach House at the University of Toronto in the Fall of 1978. "You know, I've figured out how you can put so much stock in logic and dialectic and still be so simpatico to my work," he said. "You were a musician before you joined the logic-boys!" (He was referring to my "career" as a songwriter and record producer in the late 1960s; if you have not heard of it, you're not alone, as recordings of my songs sold a nearly negative number of copies. But I had regaled Marshall and Corinne with them on more than one occasion, and had even given them a copy of my 1971 album, "Twice Upon a Rhyme.")

"Well, I've been logical a pretty long time," I replied. "And even in music, you get chords with three notes before you have four."

And, of course, wonderful chords can be had with five and six notes, too. The enduring message in triad versus tetrad, I think, is that there are no magic numbers of components in tools for social analysis. True, the tetrad resonates with profundities in four-part structures ranging from the G, A, T, Cs of DNA to the four horsemen of the Apocalypse to Aristotle's four causes to the medieval quadrivium. McLuhan was especially fond of citing that last – the study of arithmetic, geometry, music, and astronomy as the graduate (Master of Arts) curriculum at Cambridge in the thirteenth century – and the superiority it implied over the trivium (Latin grammar, rhetoric, and logic), which served as the undergraduate (Bachelor of Arts) curriculum at Cambridge at that time. (He no doubt also enjoyed the implication by pun that the triad was trivial in comparison to the tetrad.) But, of course, three-part structures have had equivalent impact in history, as witness the Trinity in Roman Catholicism and the three wishes of so many fables.

Indeed, if multiplicity is the spirit of the digital age, it encompasses the trivium as well as the quadrivium – which later joined forces to form the seven liberal arts – and much more. We might say that, as a vehicle for education not only formal but more importantly via living, the Web has obsolesced the seven liberal arts in favor of a curriculum with boundaries far less rigid, and populated by thousands of subjects, constantly under construction via the serendipity of Web links rather than departmental committees in universities.

We might call this new anti-regime: the digital arts.

DEEP RETRIEVAL AND THE DIGITAL ARTS

When we consider the properties of the various media that television is in the process of reversing into – permanence and increased control of programming via the VCR, integration of text and interaction via the Web, increased choice of programming via cable, the third dimension via holography (as yet, the least widespread of these reversals, although the first holograms were created in the late 1940s) – we find that these qualities have been long in residence in human perception and communication. The quest for permanence in human communication was first demonstrated at least tens of thousands of years ago in cave paintings. Text is newer but goes back thousands of years in hieroglyphics. Increased control of the environment, interaction among people, and the perception of the third dimension are each in their own way so fundamental to our existence as to presage and presume the very function of media.

These are the powerful points of origin, and therefore the goal, of the evolution of media, as described in my "anthropotropic" theory. They have little to do with the rear-view mirror, the lens of the proximate, immediate past, which can distract us, help us get better bearings, and often does both for us as we strive to make some initial sense out of our new media surroundings. Instead, these quintessential ingredients of human existence and communication steer the evolution of all media. They are the deepest, and thus the most important, components of what McLuhan had in mind by retrieval. They allowed him to note with typically casual but incisive sagacity that the car retrieves "the knight in shining armour" (M. McLuhan & E.

McLuhan, 1988, p. 148). And they allow us to note now that the VCR retrieves, among other things, Altamira, Lascaux, and Chauvet. The paleolithic reach and specificity of such deep retrievals is one of the great opportunities of the tetrad, and indeed not usually found in the dialectic.

Further, the very multiplicity of tetradic possibilities invites additional thought and consideration – the tetrad is thus a "cool" tool, as are all of McLuhan's insights, constructs, and metaphors. For example, why is perception of the third dimension, on the one hand the oldest of all the human traits described above (because it obviously predates humanity), also the least fulfilled or retrieved in nineteenth-century and subsequent media? The best answer I can come up with is that, although the third dimension is indisputably a component of our pre-technological environment, it is likely not as crucial as motion, sound, color, immediacy, and other aspects retrieved in the evolution of the photograph from its still, silent, black-and-white, delayed first expressions. Further, the third dimension also has been retrieved fairly well in the two-dimensional illusion of perspective, practiced in Western art since the Renaissance, and automatically engendered in the photograph. In terms of my concept of "remedial" media discussed in the previous chapter, the lack of the literal third dimension in visual media thus did not generate as much of a need for remediation as the lack of those other qualities.

And this is but one example – one aspect, of one fourth (reversal), of one tetrad, for one medium, television. It shows that the tetrads demand a lot of work – not in the sense of being sloppily unfinished, the result of a lazy, undisciplined intellect, as McLuhan's critics would have us view his project (see, again, Edmundson's 1997 characterization of McLuhan as conversing in "loose, shaggy buffaloes"), but in precisely the opposite of engaging our intellect, grabbing hold of our quest for understanding, our imagination, and propelling it on an exhilarating journey. (I cannot resist adding that I guess one must have an imagination, a zest for comprehension, in order to be so engaged. In its absence, the annoyance that McLuhan's critics evince is understandable.)

This work that the tetrad requires turns out to be a hallmark of the digital age – another feature of the reversal of television into the personal computer and the Web. Television requires no work at all. One just sits back and watches it. Not that this is necessarily bad – to the contrary, as I pointed out in a paper, "The Benefits of Watching Television" (1980), sometimes it is good to have a medium at our disposal that is so undemanding (see also McGrath, 1997, for the benefits of personal "downtime" with television).

But this also tends to reduce the impact of television as an educational tool (although a viewer can certainly be intellectually intrigued, even inspired, by a particular television program – another example, as discussed in Chapters 3 and 9, of how "the medium is the message" is best understood with "medium" not being monolithic in its effects).

In contrast, personal computers require some degree of training to be used well, as does the Web. The training can be self-administered – and I have long observed that the best teacher of any tasks to be performed with the personal computer is a real problem or job that the would-be trainee needs solved or accomplished (this is certainly the best way to learn the Web) – but it is training nonetheless. And its necessity marks a break-point, a reversal, not only of the modus operandus of television, but of radio, motion pictures, the viewing of photography, and indeed all media after the telegraph and of course the book itself (which requires literacy).

If our traditional liberal arts were designed to make us fully productive citizens in a world moved by literacy, we might say that the digital arts – not only what is on the Web but the training and knowledge to know how to find it and use it – are necessary to make us fully productive citizens in at least the first years of the new millennium.

This new digital emphasis not only on knowing but in knowing how to know – which suggests that the most fundamental form of knowing is doing – is consistent with the educational philosophy of John Dewey, which, apropos his name, went so far as to equate knowing with doing. It is also consistent with McLuhan's call for "cool" teachers – facilitators, not lecturers, who elicit student participation – and theorists from Dewey to Montessori to Piaget who have promoted the virtues of the active learner (see Chapter 9 in this book).

Online education thus becomes education about how to learn online, as much as it is education in any specified subject such as history or philosophy, in this reversal of passive, lecture-and-book based education that parallels the reversal of television into the Web. Indeed, Connected Education always explained to its students, especially in the early days when online education – online anything – needed explanation, that to take an online course was intrinsically to take a course in two things: not only the advertised subject of the course, but a course in how to learn, and more generally to conduct oneself, online. In as much as the enormous growth of the Web has made it easily available to anyone with a computer, with access therein given to information the range and depth and specificity of which can rival that available in most classrooms and even university libraries, we might be moving into an era in which formal education will be obsolesced in many of its purposes and bailiwicks, except insofar as it grants officially-sanctioned degrees. You do not have to go to school to learn how to learn online, nor do you need to sit in a classroom to learn many subjects. And as these reversals of school bells clang into greater coherence, as the discrepancy between diplomas and learning becomes increasingly clear, we might see the value of degrees decline as well. Heads of academic programs are already noting with concern (see Bronner, 1998) that students in computer science are dropping out to pursue lucrative careers, as Web designers and the like, in which talent not diploma is correctly seen as the crucial prerequisite. Bill Gates, after all, has no college degree.

But to be true to McLuhan's laws of media, we need to consider the digital age not only as the terminus of tetrads centered on television, education, books, and other vehicles in the Parthenon of the twentieth century, but as a launching pad, a reaction point, for hypothetical ages to follow.

We have thus far been considering in this book the early days of the digital age, and what McLuhan's insights can teach us about developments we already see all around us.

We turn in the next section to the digital age when it has reached the limits of its potential, the end of its cycle, and inquire what it might flip into as it approaches that point.

REVERSAL OF THE DIGITAL AGE

A key feature of the tetrad, as we have seen above, is that reversal is not the complete antithesis or opposite of what comes before. It is thus not as black to white or Lévi-Straussian bi-polar opposites, nor as the Hegelian dialectic in that regard. Part of this tetradic continuity comes from the retrieval of previously obsolesced elements in reversal. Part of it comes from the new medium or effect literally incorporating aspects of the immediately preceding medium, in a sort of rear-view mirror in action. Thus, the personal computer that television reverses into retains the television screen, or a version of it, as does cable television and the VCR. Only holography dispenses with the screen, but it retains other aspects of television, notably its one-way audio-visual form of presentation.

If the digital age is characterized by people personalizing their selection of information via Windows and Web browsers, we can expect to find similar vehicles for expression of choice in the age that follows the digital, used in different ways, for different purposes, and with different results.

Paul Verhoeven's *Starship Troopers* – his 1997 movie of Robert Heinlein's 1959 novel – depicts one such possible future, vividly. The core of the story in the Heinlein novel is a war between Earth and aliens, in which Earth has already created a unified, Sparta-like society in which voluntary military service is necessary for full citizenship, and only citizens can vote. In order for the war effort to succeed, the people of Earth must be mentally mobilized to give their all – in a word, propagandized, as was last done successfully in our own Second World War, when media were much easier to control from on high. But how, given the digital age's diminishment of gatekeeping, its encouragement of education via active selection rather than passive receipt of injections of knowledge, its bursting of the oligarchy of television into a myriad of specialized programming choices over which people have much greater control – how in this decidedly anti-propagandistic digital environment of the Web can a successful world-wide propaganda campaign be launched and implemented?

Heinlein, writing in 1959, of course did not have to face that problem. But Verhoeven, making a movie in the 1990s, most certainly did. His solution touches one of the crucial possible reversal points of the Web: the ambiguity, as far as user control is concerned, of the hypertext link. In the Web's current incarnation, people of course control the choices of others by deciding what links to include in the pages they produce; the Web surfer can only choose from links already present. But the choices are so many, and so unpredictable, that the surfer in effect has extraordinary freedom in selection of information – undoubtedly far more, likely by a combined magnitude of millions, than can be found on any television screen, certainly a screen in the 1950s.

But what happens when the hypertext links are deliberately limited, programmed to give just the information that a government wants its people to see, and no more?

Verhoeven portrays this possibility brilliantly. In a future Web in which information is presumably available on many possible subjects, the government can break in with live coverage whenever it chooses. But the pre-recorded information is itself carefully programmed and structured. A viewer sees a short clip on some aspect of the war, or the alien threat to Earth, and a typical Windows-like ribbon of further options, links to additional information, appears on the top of the screen. A deep, inviting, newscaster-like voice inquires, "Want to learn more?" But when the viewer clicks on the appropriate panel, and another short clip on the advertised subject is displayed, it turns out to be just another piece of propaganda.

Indeed, all of these pre-selected options, although conveying somewhat different information – the nature of the aliens ("know your foe"), the heroic war effort on the home front, etc. – are all created by the same slick propaganda machine. And to add insult to injury, some of the clips – presumably those portraying extreme violence – contain sections that are blocked out, or "Censored." It becomes clear – to the viewer of the movie, at least – that the only real choice those future Web browsers have is clicking on Exit, and ending that particular session of propaganda altogether.

Thus, in this scenario, the triumph of choice on our Web reverses into the illusion of choice on a future Web-like medium. Consistent with the continuity of aspects in all four sectors of the tetrad, we might well then inquire to what extent even our current levels of choice on the Web are illusory, in practice not much more than selection of pre-determined options. This is where the tetrad lends support to those who fear that a huge corporation such as Microsoft might well be gaining too much control over our digital environment. Certainly, on the basis of the tetrad, the potential for such a resurgence of centralized control, at the expense of our control, is there – indeed, the tetrad in part is designed to call attention to the seeds of reversal inherent in every medium and its effects. I am inclined to see the more likely carrier of that reversal as the government (hence, what one reviewer called my "scathing indictment of the Communications Decency Act" in *The Soft Edge*; see Levinson, 1997b, and Aufderheide, 1997, for the review) – as apparently does Verhoeven – in contrast to those who see the government as providing a defense against corporate reversals of the free Web. But we all agree that the danger is present – emanating from both Microsoft and the government, who each see the danger in the other (see Levinson, 1998b, for more – the government always carries the greater danger, because it carries guns).

But this very debate about who spearheads the danger calls attention to another, equally crucial aspect of reversal: at any given present time, such as our own, reversals in the future are potential, hypothetical, not yet real. Of the four laws of media applied to a new medium, only reversal has not yet taken place (obviously, when the tetrad is applied to an earlier medium, such as radio, we can already know its reversals). Thus, a specific, projected reversal can not only be discussed – as we can with the other three effects – but it can be specified, controlled.

Can it be prevented?

I can argue as I have in this book and elsewhere that the government not corporate power is the main threat to freedom of choice in the future (as

Jefferson saw, and I would say government has always been), but what impact can and will such an argument have? I have resisted McLuhan's stance of making no value judgements, I have disagreed with his frequent depiction of technology as in the superior cause-and-effect position to humans, precisely because I think that human rationality and evaluation *can* have a positive effect on the future – can steer reversals in this specific direction or that.

To accomplish this, we must indeed begin with value judgements. Propaganda may indeed be ubiquitous and unavoidable, as both McLuhan and Jacques Ellul (e.g., 1965/1973) saw, but that does not mean we have to embrace it equally on all occasions, or allow all of our media to become better equipped to present it. We can start in the digital age with an ethical imperative that control of information by disparate individuals is better than its control by central authorities. Propaganda in heightened form may even be needed on some occasions – as in the Second World War, or in our future war with the insect aliens of Klendathu in *Starship Troopers*. But we can recognize that in such instances we nonetheless are playing with fire, and seek better means to control it.

In our concluding section, we thus go beyond McLuhan – in effect, reverse his depiction of humans in the lap of technology – and consider how we can enhance our control of the future. Of course, since reversal always contains elements of the system from which it arose, such a discussion is as much a consideration of elements within McLuhan's work as those beyond it.

REVERSAL OF MEDIA DETERMINISM

Although McLuhan was decidedly no media determinist in the Marxist sense of seeing an inevitable future arising out of neatly explained history, he nonetheless drew his most striking examples of media and their effects in terms of how they manipulate humans without our concurrence, often even without our awareness. Who is aware when reading text on a page that the experience is flattening the multi-dimensionality of acoustic space and the world prior to the alphabet? What television viewer (other than a media theorist) ever considers that the light coming through the screen is reading us, involving us, in a way that light bouncing off the screens of motion pictures in theaters and paintings on walls cannot? Who logs on to the Web with the deliberate intention of being part of a new, interactive global village that is obsolescing the voyeuristic village engendered by television?

Not only McLuhan's works, but perforce this book as well, have gleefully trafficked in such examples for their shock value – as wake-up calls to those unaware of the profound power of media in our and every society. They are the stock-in-trade of media theorists.

But when McLuhan says that without radio there would have been no Hitler, that Nixon lost the election in 1960 because he was too hot for the new medium of television – when I began my historical account of media in *The Soft Edge* (1997b) with a discussion of how Ikhnaton's monotheism was defeated by the absence of an alphabet – we are only telling a part of the story. We are advertising the part that we think will most attract attention.

But even a moment's reflection should disclose that of course there is another part – an aspect of the human/technological relationship that says that we can do something about our inventions, refine them, guide them, to perform in ways that suit our sensibilities and needs rather than reform them. For what is the point of shocking anyone about the impact of a medium, if there is no hope of doing anything about that impact? If McLuhan sought to rouse us from our numbness at the effects of our media, surely that was because he thought that we might be in a position to continue the effects that we liked, and discontinue or at least diminish those that we did not, after our awakening.

The potentiality rather than the actuality of reversals of current media, together with the multiplicity of effects at all four nodes of the tetrad, show that this presumption of human refinement and control of technology was hardwired by McLuhan into his laws of the media. We might say that the fourth law of media – reversal – implies, perhaps insists, that humans take an active role. In that sense, this fourth law is unlike Aristotle's Fourth or Final Cause: it is an inherent destiny, an ultimate point of destination, that is ever up for grabs.

The reversal of determinism began with the arrival of life itself. Unlike inorganic reactions, the results of which are almost as predictable as two plus two equals four, living processes are animated by dollops of unpredictability. On the individual level, this unpredictability can of course lead to death as well as success; for life as a whole, this noise in determinism serves as a source of novelty via mutation, and is thus one of the cutting edges of evolution.

When that evolution gave rise to human intelligence, determinism suffered another reversal, as profound as that which attended the emergence of open-programmed life. To imagine is to disperse to infinity the prospect of a single, unavoidable result. To embody those imaginings into tangible technology is to greatly constrict that field of possibilities – for physical things are less easily wrought than ideas – but even a handful of new technologies, even just two, breaks the spell of a single, inevitable outcome.

Remedial media demonstrate this reversal of determinism for specific technologies. Rather than endure the impact of the Peeping Tom, we invented the window shade; rather than stand idly by as our favorite images flew instantly off of our television screens, we invented the VCR; rather than labor under the burden of all written words fixed inextricably to paper from the moment of their conception, we invented the word processor. Seen from afar, these reversals no doubt can be counted as automatic, inevitable flips of the media as the window, television, and writing encountered some outer limit in their function. But in actual fact, they were reversals instigated and implemented by the deliberate application of human reason. Such human initiation and accomplishment does not negate – or reverse – the process of reversal, but rather informs, directs, and transforms it, much as the emergence of humans on this world has transformed the technological enterprise already in evidence by ants, birds, beavers, and indeed all of life in one way or another.

The Internet and our current digital age that it embodies, demonstrates, facilitates, and leads is a remedial medium writ large – a reversal of inadequacies of television, books, newspapers, education, work patterns, and almost every medium and its effect that has come before. Many of these redressings were not as deliberate or intended as the VCR's cure of television's ephemerality. But their concentration in the new millennium, the combination of their assistance for such a diversity of problems with prior media, is surely no coincidence. The very speed-up and ease of communication made possible by digital media has reduced the difference between media deliberately invented to remedy a problem and media whose unintended consequences speak to a problem at hand: in the grasp of rationality enhanced by digital communication, all media become ready remedies.

And herein resides both the ambiguity of our future, and the optimism I hold for it. On the one hand, the very velocity of media evolution that has brought us to this promising juncture could pitch us into a reversal in which choice and control by individuals become trappings for much more insidious forms of gatekeeping, in which the gates are indeed open but the territory beyond is already determined. Hypertext links that connect to pages even subtly disfigured by government censorship are not windows but mirrors – fancy replays of what, at best, a small group of elected officials think we should and should not see.

On the other hand, the very velocity that has brought us to this point has brought to the surface an opportunity for individual choice and rational direction the likes of which the world has never seen. In comparison, Jefferson's Age of Reason and the citizenship it conferred only to wealthy, white males – the democracy in Ancient Greece and its segregation of barbarians – are but seedlings to now be retrieved and developed at last to their fullest.

In the new digital global village, there are no barbarians. As citizens of this new age, we have unprecedented, though of course not unlimited, power to stop reversals that seem not in our best interests, or at very least slow their advancement in favor of preservation or development of media environments that we prefer.

Thus McLuhan, in his life's work in general and his culminating laws of media in particular, may have laid bare a dynamic of media and their irresistible, unintended consequences that is transforming itself in our digital age into a dynamic of increased and enlightened human control. The increase is due to digital inventions. The enlightenment is due in no small part to McLuhan.

Is that what he intended all along?

Or is the digital undoing of media determinism, and the explication and implicit prediction of all of this in the work of Marshall McLuhan, the biggest unintended consequence of all?

Do not wait for me or anyone else to tell you.

Read McLuhan, read books and essays about his work, reread this book, and decide for yourself…

BIBLIOGRAPHY

Advertisement (1998) for *Interactive Excellence* by E. Schlossberg, *The New York Times*, 23 July: G8.

Agassi, J. (1968) *The Continuing Revolution*, New York: McGraw-Hill.

—— (1982) "In search of rationality – a personal report," in P. Levinson (ed.) (1982).

Annie Hall (1973) Motion picture directed by Woody Allen.

Asimov, I. (1945) "The mule" (concluding part 2) *Astounding Science Fiction*, December: 60–97, 148–68.

—— (1951) *Foundation*, New York: Gnome.

—— (1952) *Foundation and Empire*, New York: Gnome.

—— (1953) *Second Foundation*, New York: Gnome.

Aufderheide, P. (1997) Review of *The Soft Edge: A Natural History and Future of the Information Revolution* by P. Levinson, *In These Times*, 19 October: 32–3.

Barlow, J. P. (1994) "The economy of ideas," *WIRED*, March: 84–90, 126–9.

Bazin, A. (1967) *What Is Cinema?*, trans. H. Gray, Berkeley, CA: University of California Press.

Beatty, J. (1998) "A capital life," Review of *Titan: The Life of John D. Rockefeller, Sr.* by R. Chernow, *The New York Times Book Review*, 17 May: 10–11.

Bedazzled (1967) Motion picture directed by Stanley Donen.

Bell, D. (1975) "Technology, nature, and society," in *The Frontiers of Knowledge*, Garden City, NY: Doubleday.

Benzon, W. (1993) "The United States of the blues: On the crossing of African and European cultures in the 20th century," *Journal of Social and Evolutionary Systems*, 16, 4: 401–38.

Bester, A. (1957) *The Stars My Destination*, New York: Signet.

Birkerts, S. (1994) *Gutenberg Elegies: The Fate of Reading in an Electronic Age*, Boston: Faber and Faber.

Bliss, M. (1988) "False prophet," Review of *Letters of Marshall McLuhan*, selected and edited by M. Molinaro, C. McLuhan and W. Toye, *Saturday Night*, May: 59–60, 62.

Blisset, W. (1958) "Explorations," *Canadian Forum*, August.

Brand, S. (1987) *The Media Lab*, New York: Viking.

—— (ed.) (1985) *The Whole Earth Software Catalog for 1986*, San Francisco, CA: Point.

Broder, J. (1998) "Gore to announce 'Electronic Bill of Rights' aimed at privacy," *The New York Times*, 14 May.

Bronner, E. (1998) "Voracious computers are siphoning talent from academia," *The New York Times*, 25 June: A1, 14.

Brooks, J. (1976) *Telephone: The First Hundred Years*, New York: Harper & Row.

Brooks, T. and Marsh, E. (1979) *The Complete Directory to Prime Time Network TV Shows, 1946–present*, New York: Ballantine.

Bush, V. (1945) "As we may think," *The Atlantic Monthly*, July: 101–8.

Butler, S. (1878/1910) *Life and Habit*, New York: Dutton.

Butterfield, F. (1997) "Crime fighting's about-face," *The New York Times*, 19 January, sec. 4: 1.

Campbell, D. T. (1974a) "Evolutionary epistemology," in P. Schilpp (ed.) *The Philosophy of Karl Popper*, La Salle, IL: Open Court.

—— (1974b) "Unjustified variation and selective retention in scientific discovery," in F. J. Ayala and T. Dobzhansky (eds.) *Studies in the Philosophy of Biology*, Berkeley, CA: University of California Press.

Capra, F. (1975) *The Tao of Physics*, New York: Bantam.

Carpenter, E. (1972/1973) *Oh, What a Blow that Phantom Gave Me!* New York: Bantam.

—— and McLuhan, M. (eds.) (1960) *Explorations in Communication*, Boston: Beacon.

Carson, R. (1962) *Silent Spring*, Boston: Houghton Mifflin.

Cohn, D. L. (1951) Review of *The Mechanical Bride* by M. McLuhan, *The New York Times*, 21 October.

Curtis, J. (1978) *Culture as Polyphony: An Essay on the Nature of Paradigms*, Columbia, MO: University of Missouri Press.

—— (1987) *Rock Eras: Interpretations of Music and Society, 1954–1984*, Bowling Green, OH: Bowling Green State University Popular Press.

Cziko, G. and Campbell, D. T. (1990) "Comprehensive bibliography: Evolutionary epistemology," *Journal of Social and Biological Structures*, 13, 1: 41–82.

Dawkins, R. (1976) *The Selfish Gene*, New York: Oxford University Press.

Dennis the Menace (1993) Motion picture directed by Nick Castle.

Dizard, W., Jr. (1997) *Old Media, New Media*, 2nd ed., White Plains, NY: Longman.

Dunlap, O. E., Jr. (1951) *Radio and Television Almanac*, New York: Harper & Bros.

Dyson, E. (1997) *Release 2.0: A Design for Living in the Digital Age*, New York: Broadway.

Ebert, R. (1995) Review of *Theremin: An Electronic Odyssey*, *Chicago Sun-Times*, 15 December.

Edmundson, M. (1997) Review of *Marshall McLuhan: Escape Into Understanding: A Biography* by W. T. Gordon, *The New York Times Book Review*, 2 November: 38.

Ellul, J. (1965/1973) *Propaganda: The Formation of Men's Attitudes*, trans. K. Kellen and J. Lerner, New York: Vintage.

Engdahl, S. (1990) "The mythic role of space fiction," *Journal of Social and Evolutionary Systems*, 13, 4: 289–96.

Ferrell, K. (1996) Personal conversation about *Omni* magazine with its former editor, White Plains, NY, 12 July.

Forster, E.M. (1951) *Two Cheers for Democracy*, New York: Harcourt, Brace & World.

Freud, S. (1930) *Civilization and its Discontents*, trans. J. Riviere, New York: Cape and Smith.

From the Earth to the Moon (1998) Cable-tv series, HBO-TV, produced by Tom Hanks, Part 12, 10 May.

Fromm, E. (1941) *Escape from Freedom*, New York: Rinehart.

The Fugitive (1993) Motion picture directed by Andrew Davis.

Genie (1991–1998) (formerly: General Electric Network for Information Exchange) Online discussion in the Science Fiction Round Table.

Gibson, C. R. (n.d.) *The Wonders of Modern Electricity*, Philadelphia, PA: McKay.

Gordon, W. T. (1997) *Marshall McLuhan: Escape Into Understanding: A Biography*, New York: Basic Books.

Gore, A. (1998) Commencement address at New York University, 14 May.

Grave, W. W. (1954) "Cambridge University," *Encyclopedia Britannica*, vol. 4, Chicago: Encyclopedia Britannica.

Gray, T. (1751) "Elegy written in a country churchyard," reprinted in O. Williams (ed.) *Immortal Poems of the English Language*, New York: Washington Square Press, 1952.

Greenhouse, L. (1997) "Court, 9–0, protects speech on Internet," *The New York Times*, 27 June: A1, 20.

Head, S. W. and Sterling, C. H. (1987) *Broadcasting in America*, 5th ed., Boston: Houghton Mifflin.

Heelan, P. (1983) *Space-Perception and the Philosophy of Science*, Berkeley: University of California Press.

Heim, M. (1987) *Electric Language*, New Haven, CT: Yale University Press.

Heinlein, R. (1959) *Starship Troopers*, New York: G. P. Putnam's.

Heyer, P. (1995) *Titanic Legacy*, Westport, CT: Praeger.

Hiltz, S. R. and Turoff, M. (1978) *The Network Nation: Human Communication via Computer*, Reading, MA: Addison-Wesley. Revised edition (1993) with Foreword by S. Keller, Cambridge, MA: MIT Press.

Hogarth, S. H. (1926) "Three great mistakes," *Blue Bell*, November.

Innis, H. (1950) *Empire and Communications*, Toronto: University of Toronto Press.

—— (1951) *The Bias of Communication*, Toronto: University of Toronto Press.

Josephson, M. (1959) *Edison*, New York: McGraw-Hill.

Keepnews, P. (1976) "The latest do-it-yourself fetish: computers," *New York Post*, 9 June: 47.

Kelly, P. (1997a) "Evolutionary epistemology and media evolution," *Journal of Social and Evolutionary Systems*, 20, 3: 233–52.

—— (1997b) "Self-organization in media evolution: A theoretical prelude to the Internet," M.A. thesis, The New School for Social Research.

—— (1997c) Unpublished Letter to the Editor of *The New York Times* about M. Edmundson (1997), 5 November.

Leave it to Beaver (1997) Motion picture directed by Andy Cadiff.

Lehmann-Haupt, C. (1989) Review of *Marshall McLuhan: The Medium and the Messenger* by P. Marchand, *The New York Times*, 20 March: C17.

Levinson, P. (1976) "'Hot' and 'cool' redefined for interactive media," *Media Ecology Review*, 4, 3: 9–11.

—— (1977a) Preface to M. McLuhan (1977a).

—— (1977b) "Toy, mirror, and art: The metamorphosis of technological culture," *et cetera*, 34, 2: 151–67. Reprinted in L. Hickman and A. al-Hibri (eds.) *Technology and Human Affairs*, St. Louis, MO: C.V. Mosby, 1981. Reprinted in L. Hickman (ed.) *Philosophy, Technology, and Human Affairs*, College Station, TX: Ibis, 1985. Reprinted in L. Hickman (ed.) *Technology as a Human Affair*, New York: McGraw-Hill, 1990. Reprinted in P. Levinson (1995b).

—— (1978a) "The future of technology," Seminar with Marshall McLuhan, Centre for Culture & Technology, University of Toronto, 10 November.

—— (1978b) "Tetradic wheels of evolution," Paper presented at Tetrad Conference (1978).

—— (1979a) "Human replay: A theory of the evolution of media," Ph.D. diss., New York University.

—— (1979b) Review of *Culture as Polyphony* by J. Curtis, *Technology and Culture*, 20, 4: 835–7.

—— (1980) "Benefits of watching television," *ERIC* microfiche, #ED 233404.

—— (1981a) "McLuhan and rationality," *Journal of Communication*, 31, 3: 179–88.

—— (1981b) "McLuhan's contribution in an evolutionary context," *Educational Technology*, 22, 1: 39–46.

—— (1981c) "McLuhan's misunderstood message," Letter to the Editor, *The New York Times*, 24 February: A18.

—— (1981d) "Media evolution and the primacy of speech," *ERIC* microfiche, #ED 235510.

—— (1981e) Personal correspondence to Loretta Barrett, Doubleday editor, about publication of *Laws of Media*, 7 January.

—— (1982) "What technology can teach philosophy," in P. Levinson (ed.) (1982).

—— (1984) "The New School online," unpublished report prepared for The New School for Social Research, December. Excerpted as "Basics of computer conferencing, and thoughts on its applicability to education," in P. Levinson (1995b).

—— (1985) "Impact of personal information technologies on American education, interpersonal relationships, and business, 1985–2010," report prepared for the U.S. Army Research Institute, February. Reprinted in P. Levinson (1995b).

—— (1986) "Marshall McLuhan and computer conferencing," *IEEE Transactions of Professional Communications*, March: 9–11. Reprinted in P. Levinson (1995b).

—— (1988a) Letter to the Editor, *Saturday Night*, August: 6.

—— (1988b) *Mind at Large: Knowing in the Technological Age*, Greenwich, CT: JAI Press.

—— (1989) "Intelligent writing: The electronic liberation of text," *Technology in Society*, 11, 3: 387–400. Reprinted in M. Fraase, *Hypermedia Volume Two*, Chicago, IL: Scott, Foresman, 1990. Reprinted in P. Levinson (1995b).

—— (1990) "McLuhan's space," Essay/Review of *Marshall McLuhan: The Medium and the Messenger* by P. Marchand; *Laws of Media* by M. McLuhan and E. McLuhan; *The Global Village* by M. McLuhan and B. R. Powers; and *Letters of Marshall McLuhan* selected and

Levinson, P. (*continued*)
edited by M. Molinaro, C. McLuhan, and W. Toye, *Journal of Communication*, 40, 2: 169–73.

—— (1992) *Electronic Chronicles: Columns of the Changes in our Time*, San Francisco, CA: Anamnesis Press.

—— (1993) Review of *Marshall McLuhan: The Medium and the Messenger* by P. Marchand; *Laws of Media* by M. McLuhan and E. McLuhan; *The Global Village* by M. McLuhan and B. R. Powers; and *Letters of Marshall McLuhan* selected and edited by M. Molinaro, C. McLuhan, and W. Toye, *WIRED*, July–August: 104–5.

—— (1994a) "Burning down the house," *WIRED*, March: 76.

—— (1994b) "Telnet to the future?" *WIRED*, July: 74.

—— (1995a) "The chronology protection case," *Analog: Science Fiction and Fact*, September: 100–18. Reprinted in C. G. Waugh and M. Greenberg (eds.) *Supernatural Sleuths*, New York: ROC Books, 1996. Reprinted in *Infinite Edge*, June, 1997. Reprinted in J. Dann (ed.) *Nebula Awards 32: SFWA's Choices for the Best Science Fiction and Fantasy of the Year*, New York: Harcourt Brace, 1998.

—— (1995b) *Learning Cyberspace: Essays on the Evolution of Media and the New Education*, San Francisco, CA: Anamnesis Press.

—— (1995c) "Web of weeds," *WIRED*, November: 136.

—— (1997a) "Learning unbound: Online education and the mind's academy," *Analog: Science Fiction and Fact*, March: 48–57.

—— (1997b) *The Soft Edge: A Natural History and Future of the Information Revolution*, London and New York: Routledge.

—— (1998a) "The book on the book," *Analog: Science Fiction and Fact*, June: 24–31.

—— (1998b) "Leave Microsoft alone," *The Industry Standard*, 8 June: 36.

—— (1998c) "Way cool text through light hot wires," Paper presented at Symposium on Marshall McLuhan (1998).

Levinson, P. and Schmidt, S. (forthcoming) "From gatekeeper to matchmaker: The shape of publishing to come."

Levinson, P. (ed.) (1982) *In Pursuit of Truth: Essays on the Philosophy of Karl Popper*, Atlantic Highlands, NJ: Humanities Press.

Lippmann, W. (1927) *The Phantom Public*, New York: Macmillan.

Lohr, S. (1998a) "Microsoft fight will be waged on wide front," *The New York Times*, 20 May: A1, D4.

—— (1998b) "Software and hardball: Potential grip of Microsoft, via Internet, on many industries is at center of dispute," *The New York Times*, 14 May: A1, D3.

MacDonald, D. (1967) "He has looted all culture...," in G. E. Stearn (ed.), 1967.

Marchand, P. (1989) *Marshall McLuhan: The Medium and the Messenger*, New York: Ticknor & Fields.

McGrath, C. (1997) "Giving Saturday morning some slack," *The New York Times Magazine*, 9 November, sec. 6: 52.

McLuhan, M. (1951) *The Mechanical Bride: Folklore of Industrial Man*, New York: Vanguard.

—— (1960) "Report on project in understanding new media", typescript published by the National Association of Educational Broadcasters, US Department of Health, Education, and Welfare, Washington, DC, 30 June.

—— (1962) *The Gutenberg Galaxy*, New York: Mentor.

McLuhan, M. (*continued*)

—— (1964) *Understanding Media*, New York: Mentor. Reprint edition (1994), with an Introduction by L. Lapham, Cambridge, MA: MIT Press.

—— (1967a) "Casting my perils before swains," Preface to G. E. Stearn (ed.), 1967.

—— (1967b) *Verbi-Voco-Visual Explorations*, New York: Something Else Press.

—— (1969) "Playboy interview: Marshall McLuhan – a candid conversation with the high priest of popcult and metaphysician of media," *Playboy*, March: 53–74, 158.

—— (1970) *Culture is Our Business*, New York: Ballantine.

—— (1975) "McLuhan's laws of the media," *Technology and Culture*, January: 74–8.

—— (1976) "Inside on the outside, or the spaced-out American," *Journal of Communication*, 26, 4: 46–53.

—— (1977a) "The laws of the media," with a Preface by P. Levinson, *et cetera*, 34, 2: 173–9.

—— (1977b) Personal correspondence to Paul Levinson, 8 September.

—— (1978) "A last look at the tube," *New York*, 3 April: 45.

McLuhan, M. and Fiore, Q. (1967) *The Medium is the Massage: An Inventory of Effects*, New York: Bantam.

—— —— (1968) *War and Peace in the Global Village*, New York: Bantam.

McLuhan, M. and Hutchon, K. and McLuhan, E. (1977) *City as Classroom*, Agincourt, Ontario: Book Society of Canada.

McLuhan, M. and McLuhan, E. (1988) *Laws of Media: The New Science*, Toronto: University of Toronto Press.

McLuhan, M. and Nevitt, B. (1972) *Take Today: The Executive as Dropout*, New York: Harcourt Brace Jovanovich.

McLuhan, M. and Parker, H. (1968) *Through the Vanishing Point: Space in Poetry and Painting*, New York: Harper & Row.

—— —— (1969) *Counterblast*, New York: Harcourt, Brace & World.

McLuhan, M. and Powers, B. R. (1989) *The Global Village*, New York: Oxford University Press.

McLuhan, M. and Watson, W. (1970) *From Cliché to Archetype* , New York: Viking.

McNeill, W. (1982) *The Pursuit of Power*, Chicago, IL: University of Chicago Press.

McWilliams, P. A. (1982) *The Personal Computer Book*, Los Angeles: Prelude.

Medium Cool (1969) Motion picture directed by Haskell Wexler.

Meyrowitz, J. (1977) "The rise of 'middle region' politics," *et cetera*, 34, 2: 133–44.

—— (1985) *No Sense of Place*, New York: Oxford University Press.

Miller, J. (1971) *Marshall McLuhan*, New York: Viking.

Mission Impossible (1996) Motion picture directed by Brian de Palma.

Molinaro, M., McLuhan, C. and Toye, W. (eds.) (1987) *Letters of Marshall McLuhan*, Toronto: Oxford University Press.

Morrow, J. (1980) "Recovering from McLuhan," *AFI Education Newsletter*, 3: 1–2.

Mortimer, J. (1988) "Tedium is the message," Review of *Letters of Marshall McLuhan*, selected and edited by M. Molinaro, C. McLuhan and W. Toye, *The Sunday Times* (London), 13 March: G1.

MSNBC-TV (1997) *Time and Again*, Rebroadcast of Frank McGee's commentary on Khrushchev–Nixon "Kitchen" debates, originally broadcast 24 July 1959, rebroadcast 24 July 1997.

Nee, E. (1998) "Surf's up," *Forbes*, 27 July: 106–13.

Nelson, T. (1980/1990) *Literary Machines*, Sausalito, CA: Mindful Press.

Office of the Independent Counsel (1998) "Referral to the United States House of Representatives," 9 September.

Orwell, G. (1948/1949) *1984*, New York: Harcourt Brace.

Perkinson, H. (1982) "Education and learning from our mistakes," in P. Levinson (ed.) (1982).

Peterson, I. (1998) "Rethinking ink," *Science News*, 20 June: 396–7.

Plato. *Phaedrus*, in B. Jowett (trans.) (n.d.) *The Dialogues of Plato*, New York: Scribner, Armstrong.

Popper, K. R. (1945) *The Open Society and its Enemies*, London: George Routledge & Sons.

—— (1957) *The Poverty of Historicism*, London: Routledge & Kegan Paul.

—— (1972) *Objective Knowledge: An Evolutionary Approach*, London: Oxford.

—— (1974) "Autobiography," in P. Schilpp (ed.) *The Philosophy of Karl Popper*, La Salle, IL: Open Court.

Postman, N. (1985) *Amusing Ourselves to Death*, New York: Viking.

—— (1992) *Technopoly: The Surrender of our Culture to Technology*, New York: Knopf.

—— (1994) "John Culkin Memorial Talk," New School for Social Research, New York City, 16 February.

—— (1998) "Six questions about media," Paper presented at Symposium on Marshall McLuhan (1998).

Pulp Fiction (1994) Motion picture directed by Quentin Tarantino.

Reservoir Dogs (1992) Motion picture directed by Quentin Tarantino.

Rheingold, H. (1993) *The Virtual Community*, Reading, MA: Addison-Wesley.

Richards, I. A. (1929) *Practical Criticism*, London: K. Paul, Trench, Trubner.

Rosenthal, R. (ed.) (1968) *McLuhan: Pro and Con*, Baltimore: Penguin.

Sawyer, R. (1990/1996) "WordStar: A writer's word processor," available on http://www.sfwriter.com/wordstar.htm.

Schmidt, S. (1989) "Pure art and electronics," *Analog: Science Fiction and Science Fact*, September: 4–12.

Schreiber, F. (1953) "The battle against print," *The Freeman*, 20 April.

Schwartz, T. (1973) *The Responsive Chord*, Garden City, NY: Anchor/Doubleday.

Shannon, C. and Weaver, W. (1949) *The Mathematical Theory of Communication*, Urbana, IL: University of Illinois Press.

Skagestad, P. (1993) "Thinking with machines: Intelligence augmentation, evolutionary epistemology, and semiotic," *Journal of Social and Evolutionary Systems*, 16, 2: 157–80.

—— (1996) "The mind's machines: the Turing machine, the Memex, and the personal computer," *Semiotica*, 111, 3/4: 217–43.

Sokolov, R. (1979) Review of *The Printing Press as an Agent of Change* by E. L. Eisenstein, *The New York Times Book Review*, 25 March: 16.

Specter, M. (1998) "Europe, bucking trend in U.S., blocks genetically altered food," *The New York Times*, 20 July: A1, 8.

Starship Troopers (1997) Motion picture directed by Paul Verhoeven.

Stearn, G. E. (ed.) (1967) *McLuhan: Hot & Cool*, New York: Dial.

Stevens, H. (1987) "Electronic organization and expert networks: beyond electronic mail and computer conferencing," in *Proceedings of IEEE Conference on Management and Tech-*

nology: Management of Evolving Systems, New York: Institute of Electrical and Electronics Engineers.

Strate, L. and Wachtel, E. (eds.) (forthcoming) *The Legacy of McLuhan*, New York: Fordham University Press.

Symposium on Marshall McLuhan (1998) Fordham University, New York City, 27–8 March.

Tedford, T. L. (1985) *Freedom of Speech in the United States*, New York: Random House.

Tetrad Conference with Marshall McLuhan (1978) Fairleigh Dickinson University, Teaneck/Hackensack, New Jersey, 10 March.

Theall, D. (1971) *The Medium is the Rear-View Mirror: Understanding McLuhan*, Montreal: McGill-Queens University Press.

Theremin: An Electronic Odyssey (1995) Motion picture (documentary) directed by Steven M. Martin.

Tipler, F. (1994) *The Physics of Immortality*, New York: Doubleday.

Titanic (1997) Motion picture directed by James Cameron.

Titanic: Secrets Revealed (1998) Television documentary, produced by Tribune Entertainment, broadcast on WPIX-TV (New York City), 25 June.

Turoff, M. (1985) "Information and value: the internal information marketplace," *Journal of Technological Forecasting and Social Change*, July, 27, 4: 257–373.

Vacco, D. (1998) Comments on CNN (Cable News Network) TV, 19 May.

van Gelder, L. (1985) "The strange case of the electronic lover," *Ms.*, October: 94ff.

Wachtel, E. (1977/1978) "The influence of the window on Western art and vision," *The Structurist*, 17/18: 4–10.

—— (1997) "McLuhan for beginners," comment entered on the Media Ecology listserv on the Internet, 3 November.

Will, G. (1998) Review of *The Last Patrician: Bobby Kennedy and the End of American Aristocracy* by M. K. Beran, *The New York Times Book Review*, 24 May: 5–6.

Winner, L. (1977) *Autonomous Technology*, Cambridge, MA: MIT Press.

Wolfe, T. (1965/1967) "What if he is right?" in G. E. Stearn (ed.) (1967).

Wolff, M. (1998) "Louis the un-Wired," *The Industry Standard*, 18 May: 10.

Wynn, M. (1977) *The Plug-In Drug*, New York: Viking.

INDEX